MAN AND THE ATOM

THE USES OF NUCLEAR ENERGY

THE WORLD OF SCIENCE LIBRARY

GENERAL EDITOR: ROBIN CLARKE

MAN AND THE ATOM
THE USES OF NUCLEAR ENERGY

Frank Barnaby

 THAMES AND HUDSON LONDON

To Sandra

Set by Filmtype Services Ltd., Scarborough
Printed in Holland by Smeets Lithographers, Weert
Bound in Holland by Proost en Brandt NV, Amsterdam
0 500 08009 7 Clothbound
0 500 10010 1 Paperback

CONTENTS

INTRODUCTION

Mankind stands on the threshold of the nuclear age – an age in which the power of the atomic nucleus will be utilized in a host of different ways, perhaps most dramatically in helping to satisfy the world's rapidly growing demand for energy. Available energy is at present produced mainly by human and animal muscle, and by combustion; as yet, only a small fraction comes from nuclear sources. In the developing countries most energy is provided by animal power and the combustion of wood and dung. But in the industrialized countries the most important source is the combustion of the fossil fuels coal, oil and natural gas. Combustion is a chemical reaction involving the rapid combination of oxygen in the air with hydrogen and carbon in fuels; energy is produced as a consequence of the rearrangement of the atoms of the hydrogen, carbon and oxygen taking part in the reaction, resulting in the formation of water and carbon dioxide. Only the outer electrons of the atoms are involved in this process; the atomic nuclei are wholly unaffected by the combustion reaction. In contrast, nuclear energy arises from events which take place inside the nuclei of atoms.

It has been said that energy is the lifeblood of nations. Without it no wheels would turn and there would be no heat; most human activity would grind to a halt. Economic progress too is obviously dependent on the availability of abundant energy – standards of living

An indication of New York's enormous demands for energy is given by these myriad lights. The disastrous effects of a power shortage on an advanced technological society stress the importance of adequate reserves of electricity-generating capacity as can be provided by nuclear power

prove to be highest in those societies where the per capita utilization of energy is largest. As Edward S. Mason of Harvard University has put it: 'Large differences in income are associated with large differences in energy intake, and we may take it for granted that no country at this stage of history can enjoy a high per capita income without becoming an extensive consumer of energy.'

There has been a continuous shift in the relative importance of various energy sources. Ever since muscle power ceased to be the sole source of energy, each new fuel that has become important has gradually lost its share of the market as new sources have appeared and expanded. However dominating its position, each fuel has been superseded in turn by new discoveries and new technologies. Thus wood was superseded as the major world fuel by coal in about 1880 and coal in turn has just been overtaken by oil. And, eventually, oil will be superseded by nuclear fuel.

The consumption of energy is increasing at a staggering rate. In 1980, the world will consume twice as much energy as it did in 1960 and probably five times as much by the year 2000. But the rate of increase is likely to differ between regions. In 2000, the countries of North America, Western Europe and Oceania will use about 45 per cent of the world total, compared with 60 per cent in 1960. And the share of the Soviet Union and Eastern Europe will remain constant at about 20 per cent. But that of the rest of the world will rise from 20 per cent in 1960 to 35 per cent in 2000. This increase will be mainly due to the industrialization of the world's developing regions.

Nuclear energy will, by producing electricity – the most versatile source of energy – play an essential role in meeting these massive energy needs. But initially it will make its greatest impact in the more advanced countries. In the United States, for example, the total installed electricity-generating capacity is at present about 300 million kilowatts, and it is predicted that this must be more than tripled by 1990 to meet the then-anticipated demand for one billion (a thousand

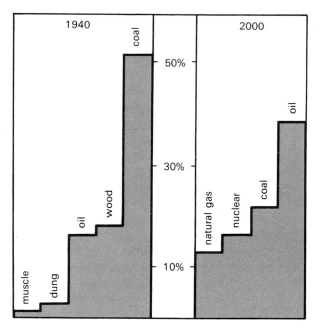

million, 10^9) kilowatts. If these amounts of electricity were produced by oil-fired power stations, about 8 million barrels of oil would be required per day at present and about 25 million in 1990. These soaring demands for electricity can be met only by the increasing use of nuclear power – conventional methods would be inadequate. And towards the end of the century the developing countries will increasingly turn to nuclear energy to provide power for their new industries.

But the nuclear revolution will by no means end at the production of energy. Nuclear energy will have many other essential uses. One will be the production of fresh water from salt water. Both the developed and the developing countries have urgent and increasing needs for cheap and plentiful water. By the 1980s, nuclear desalting will be sufficiently cheap for municipal supplies and agricultural use, especially in arid regions.

The use of radioactive substances, in agriculture, medicine and industry, will contribute as much to our

The growing importance of nuclear fuel can be seen here to coincide with a decline in the use of fossil fuels. Soon after the year 2000 nuclear fuel will be the major energy source although some countries will still be relying on muscle power and the burning of dung. A conversion factor has been used to make the comparison possible

living standards as will nuclear power. They have a vast and growing range of applications and have already become standard tools in most branches of applied research. And their use has resulted in improved food supplies, better health, and new industrial products.

Nuclear-propelled warships have revolutionized naval strategy and interest in nuclear merchant ships for the rapid transport of cargo and passengers is growing. Nuclear-powered space-craft are being developed as the only means of transporting man to the other planets of our solar system.

If a large amount of energy is liberated within a very short time and in a small space, an explosion results. And, in a nuclear explosion, the sudden uncontrolled release of a tremendous amount of nuclear energy produces a devastating effect. Thus, the peaceful uses of nuclear explosives offer promise for civil engineering works in which very large quantities of earth must be moved and for obtaining otherwise irrecoverable natural resources or excavating underground storage spaces.

Nuclear energy is already playing a vital role in the lives of all the citizens of the industrialized countries and, in the coming years, it will increasingly contribute to the health and prosperity of all the earth's peoples. And extremely large supplies of very cheap energy produced by nuclear power will eventually be used on a wide scale as raw material for a variety of new industrial and chemical processes. This will promote a new industrial revolution with far-ranging social implications. For example, thriving communities will be established where today there are just empty deserts.

It is clear that nuclear energy is laden with potential for good. But, as with many other technologies, there is another side to the coin – it is fraught with an equal or greater potential for evil. Nuclear explosives have become the basis for weapons of greater destructive power than any ever before devised by man. In fact, the cumulative power of the nuclear weapons in existing arsenals could eliminate human life on this planet thousands of times over. Just one of the larger nuclear weapons, for example, has a destructive power far greater than that of all the conventional explosives used in all the wars in history put together.

The awesome power of these weapons is indicated by the destruction wrought in Hiroshima and Nagasaki by the explosion of relatively small nuclear weapons. A single weapon was detonated at a height of about 550 metres above each of the cities. Within seconds, a ball of fire formed. This expanded rapidly and developed into the familiar mushroom-shaped cloud on a column of black smoke. The heat from the fire-ball started thousands of fires – some at distances of one-and-a-half kilometres. Buildings were completely destroyed within a circular area of radius two-and-a-half kilometres and severe damage to houses occurred as far out as eight kilometres. The explosion over Hiroshima led to a terrifying fire-storm which lasted for about six hours and which burnt out an area of twelve square kilometres. The fire-fighting services, decimated by the blast, could not cope with the emergency; medical services were crippled; water and food supplies were disrupted; and electricity was cut off. These difficulties severely exacerbated the psychological effects on those who survived the disasters.

But 78,000 people in Hiroshima and 27,000 in Nagasaki did not survive; all that remained of some were charred shadows on the pavement. In addition, 84,000 were injured in Hiroshima and 41,000 in Nagasaki; and there were many thousands more missing in both cities. But the tragedy did not end there. Many of the survivors have since suffered from long-term effects of their exposure to radiation. The incidence of leukaemia among them, for example, still remains much higher than in the population of the rest of Japan, particularly among those who were children at the time of the explosion. The incidence is fifty times greater among people who were within one kilometre of the explosion than among those who were further away. And there is also found to be an increased incidence of other kinds of malignant cancer, particularly cancer of the thyroid gland. Possibly the most dread effects of all are the genetic mutations, the overwhelming majority of which are deleterious. These can be induced by the radiations from nuclear explosives and lead to serious physical and mental disabilities in future generations. Only time will tell the full extent of the genetic damage done by the first two nuclear disasters.

Modern nuclear weapons are typically equivalent in explosive power to about 1,000,000 tons of TNT – 50 times more powerful than the weapons dropped in anger over Japan. The scale of death, injury and physical destruction which would follow the explosion of such a weapon over a city would be so great that the consequences are unimaginable. As an organized unit the city would cease to exist. Yet in recent years even larger weapons have been experimentally exploded, with explosive powers of up to 3,000 times greater than the first bombs. These fearful new weapons are capable of destroying hundreds or even thousands of square kilometres by blast and fire.

The effects of a full-scale nuclear war would not be confined to the combatants. Radioactive fall-out would travel through the atmosphere in a vast deadly cloud and neighbouring, and even remote, countries would be exposed to the effects of radiation. And, if a large number of nuclear weapons were used, the entire world population could be endangered by radiation from fall-out particles deposited on the ground and through the ingestion of contaminated foods. The risk of nuclear war will remain for as long as nuclear weapons exist.

Some radioactivity has already been spread across the world as a result of the testing of nuclear weapons. Tests have taken place in the atmosphere, underground and undersea. Despite the spirit of the 1963 partial test-ban treaty, which prohibited tests in the atmosphere, in outer space, and underwater, the United States and the Soviet Union have continued their testing programmes without abatement underground. And France and China, who have not signed the treaty, continue their tests in the atmosphere. It appears that the annual average number of nuclear tests by all nations between 1950 and 1963 was 40. Since 1963 the average has in fact increased to 46.

Estimates of the number of people affected by fall-out radiation from weapons testing have varied enormously (some scientists have claimed that it is many hundreds of thousands). The amount of fall-out in the atmosphere has considerably decreased since the test-ban, but it is possible that underground water sources may become radioactively contaminated as a result of underground tests – a danger that will grow as the size of the weapons increases.

Accidents involving nuclear weapons represent another serious hazard; a nuclear war might conceivably be started if one country were accidentally to detonate a weapon on the territory of another nuclear-weapon power. There have been many accidents involving bombers carrying nuclear weapons; and nuclear-powered submarines – for example, the U.S.S. Scorpion – have been lost at sea. In January 1966, an American B-52 collided in mid-air with a refuelling tanker near Palomares, Spain. The B-52 crashed and four very large nuclear weapons separated from the aircraft. One landed intact in a riverbed; two others released radioactivity right in the middle of a populated area; and a fourth fell in the sea and was recovered about three months later after an intensive search. And two years later four nuclear weapons were lost when a B-52 crashed at Thule, Greenland.

Already it is clear that much good can be expected from nuclear energy – cheap and abundant power, fresh water, larger crops of improved quality, better health, new methods for industry, large-scale civil engineering, recovery of natural resources, cheap propulsion for ships and transport to other planets. But the world would obtain these benefits at a much faster rate if such large resources were not being channelled into the nuclear arms race and away from peaceful applications. If this arms race is allowed to continue we may never obtain them.

THE BIRTH OF THE BOMB

1

Since the beginning of the century it has been known that a large amount of energy is locked up in the nucleus of the atom. What was needed was a key to release this energy; and just such a key was found in 1932, when James Chadwick discovered the neutron – the tiny, uncharged particle found within the atomic nucleus. But, as has happened so often in science, few physicists were able to appreciate the significance of the event. One of those who did was Fritz Houterman, who stated, shortly after the discovery, that the particle might be capable of releasing the powerful forces dormant in matter. But authoritative opinion (supported by such eminent scientists as Niels Bohr, Albert Einstein and Ernest Rutherford) was ranged against this minority; and, right up to 1939, scientists were on the whole pessimistic about the practical exploitation of nuclear energy.

The event which forced them to change their minds was the discovery, by Otto Hahn and Friedrich Strassmann in Germany, that when a nucleus of uranium absorbed a neutron it would sometimes split up into two fragments. During this process – called fission, a term borrowed from biologists – some of the energy dormant in the nucleus is released. But for an accident, nuclear fission would have been discovered in 1934 instead of in 1938. And if this had happened, Hitlerite Germany might well have produced the first nuclear bomb. During the 1930s, a great deal of nuclear re-

On August 6, 1945 man's predicament was fundamentally changed when the first nuclear weapon to be used in war was detonated over Hiroshima. The main threats to survival are no longer natural events like famines and earthquakes but the actions of man himself

search was going on in Germany and Italy, and German scientists were, to an extent unusual at the time, actively engaged in military technology. Fermi and his collaborators, working in Rome, systematically bombarded one element after another with neutrons and, in 1934, they tried uranium. Neutrons are ideal projectiles with which to bombard nuclei because they carry no electric charge and therefore can easily approach and enter nuclei. Apart from neutrons, a nucleus contains protons, which are similar particles except that they carry positive electric charges. The nucleus itself is, therefore, positively charged and will repel any protons which come close to it, whereas neutrons are unaffected.

It was found that the bombardment of uranium gave rise to a new element, or perhaps even several new ones. Fermi concluded that transuranic elements heavier than uranium had been produced in the laboratory. But Ida Noddack, a German chemist, published a paper giving an alternative explanation of the results. She suggested that a nucleus of uranium, after absorbing a neutron, might break up into two fragments or, in other words, that Fermi had produced nuclear fission. But Fermi did not take this suggestion seriously; in fact, at the time he believed that fission was not possible. Meanwhile, many other workers were attempting to identify Fermi's supposed transuranic elements.

Hahn and Strassmann made an unusually thorough chemical analysis of a uranium target which had been exposed to neutrons and were surprised to find that the element barium was produced in the target by the neutron bombardment. And, at about the same time, Irène Joliot-Curie identified the element lanthanum in a similar experiment. These discoveries led Lise Meitner and Otto Frisch to suggest the theory of fission. In a letter to the scientific journal *Nature*, dated 16th January 1939, they wrote: 'It seems possible that the uranium nucleus has only small stability of form, and it may, after neutron capture, divide itself into two nuclei of roughly equal size.'

Meitner and Frisch correctly predicted that a relatively large amount of energy is liberated during fission. Even so, the fission process alone is insufficient for the practical utilization of nuclear energy. This could only be achieved if fission, which is initiated by neutrons, is also accompanied by the emission of neutrons and if these are, in turn, able to initiate further fissions in neighbouring uranium nuclei. Very shortly after the discovery of fission, Hans van Halban and his associates verified that neutrons are emitted when a uranium nucleus undergoes fission. And, provided that at least one neutron could be made to split another nucleus, a self-sustaining process or chain reaction could be produced and energy generated continuously. But in practice the chain reaction was still far away.

A major problem was related to the nature of uranium. Ordinary uranium consists of a mixture of three kinds of atom. It is now common knowledge that the nucleus of an atom contains just protons and neutrons, and that all the nuclei of a given element contain the same number of protons, the atomic number. For uranium this is 92. But one type of uranium atom contains 142 neutrons in its nucleus, a second type contains 143 and a third 146. These types (or isotopes) of uranium are referred to as U-234, U-235 and U-238; 'U' is the chemical symbol for uranium and the number represents the total number of neutrons and protons in the nucleus. In 1939, Niels Bohr predicted that a neutron can cause fission in U-238 only if its velocity exceeds a certain value. But too few of the neutrons available for sustaining the fission process have this critical velocity and a chain reaction is not possible using only U-238. Bohr also predicted that a nucleus of U-235 will undergo fission when any neutron, even one moving very slowly, collides with it, and that a chain reaction is possible using U-235. This was proved experimentally in 1940 by John Dunning and his associates at Columbia University.

Consequently, the fission process of greatest practical importance consists of the capture of a neutron by a U-235 nucleus, resulting in the formation of a nucleus

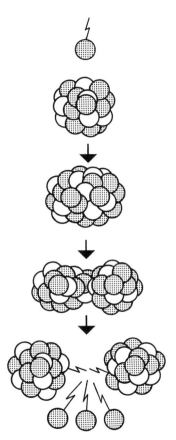

The fission process. A single neutron (shaded) enters a uranium-235 nucleus to make unstable uranium-236 which splits to make two fission fragments – nuclei of medium weight elements – and release two or three neutrons

of the isotope U-236, which rapidly splits into two fragments, nuclei of elements of medium atomic numbers called fission products. On average, about 2.5 neutrons are produced during this process, which can be represented by the equation: U-235 + neutron → U-236 → X + Y + 2.5 neutrons, where X and Y are the fission products.

An explanation of the release of fission energy can be given in terms of the famous Einstein equation. In 1905, Einstein advanced the then revolutionary concept that matter and energy are two aspects of a primal cosmic substance, rather than two entirely separate phenomena. Moreover, he claimed that matter could be converted into energy and expressed the relationship between them by the terse equation $E = mc^2$, where E represents the energy equivalence of a quantity of matter (or mass) m, and c is the velocity of light. It invariably happens that the total sum of the masses of the fission products and the fission neutrons is less than the mass of the U-236 nucleus. The energy accompanying fission is equal to this mass difference multiplied by the square of the velocity of light. Although the mass difference involved is very small, the velocity of light squared is an enormous number and, therefore, the amount of energy liberated is very large. In fact, the complete fissioning of one gram of U-235 would release about 23,000 kilowatt-hours of heat. Put in another way, one ton of uranium has roughly the same potential fuel value as 3 million tons of coal or 12 million barrels of oil.

Production of plutonium

Substances like U-235, which will undergo fission after capturing either a slow or a fast neutron, are called fissile materials, and these are of fundamental value for the utilization of nuclear energy. In experiments performed in the Berkeley Radiation Laboratory during 1941, it was found that plutonium-239 was a fissile material, a discovery which immediately aroused great interest in it.

Pu-239 is produced when U-238 nuclei absorb slow

neutrons. The U-239 nuclei so formed do not undergo fission but are ultimately transformed into Pu-239. U-239 is an example of a radioactive isotope, or radionuclide, and as such it spontaneously disintegrates or decays.

A U-239 nucleus decays by emitting an electron which is, so to speak, shot out of the nucleus like a bullet. But when this process was discovered it was not at first realized that the particle emitted was an electron, and so it was called a beta particle, a name which is still used today. The emission of a beta particle occurs when one of the neutrons in the radioactive nucleus spontaneously changes into a proton. A new element is formed by this fascinating event since the new nucleus contains an additional proton. Thus, when U-239 decays an isotope with an atomic number of 93 is produced, namely neptunium-239 (Np-239). And Np-239 is also radioactive, undergoing beta decay to produce Pu-239 (atomic number 94).

As fresh fuel elements are inserted into the front face of this plutonium production reactor at Hanford irradiated elements are discharged at the other end. The plutonium was extracted from the fuel elements by a chemical process and used for the production of nuclear weapons

Radionuclides are characterized by their half-lives – the time taken for half of a large number of the radioactive atoms to decay. Pu-239 is itself radioactive but whereas the half-lives of U-239 and Np-239 are very short (23 minutes and 2.3 days respectively) Pu-239 has the very long half-life of 24,000 years. And Pu-239 decays in a different manner to the other two radionuclides. Instead of emitting a beta particle it shoots out two protons and two neutrons tightly bound together, and this composite entity is called an alpha particle. The nucleus left after this event is U-235. But if the Pu-239 nucleus captures a neutron before it decays in this manner it will undergo fission – and therein lies its importance.

The Manhattan Project

During the early 1940s, the major concern of the United States and the Allied Nations about U-235 and Pu-239 was whether they could be used to make a nuclear weapon. To impress upon the American government the possible military consequences of the nuclear research then being carried out, Leo Szilard, a Hungarian physicist living in the United States, persuaded Einstein in October 1939 to send a letter to the President in which he recommended that the American government, although neutral at the time, should take an active interest in these matters to guard against the possibility of a surprise German nuclear weapon. Ironically, the construction of a German weapon was largely frustrated by the fact that many eminent scientists had been driven into exile by the Nazi regime, and by the lack of enthusiasm evinced by many of the nuclear scientists who remained in Germany during this period.

In contrast, the vast majority of the nuclear scientists in the allied countries were willing to devote their entire efforts to the production of the new weapon. But official support for these efforts was slow in coming. In Britain, a committee was set up under George P. Thomson to follow the work being done on nuclear fission. And in July 1941 it reported that it was 'quite

probable that the atom bomb may be manufactured before the end of the war'.

In the United States, it was not until 6 December 1941, the very day before the Japanese attack on Pearl Harbour, that the decision was taken to channel substantial resources into the nuclear weapons project. It had taken the scientists two years of frantic effort to arouse the interest of the military. In 1942, it was agreed that the British and American nuclear research teams should be concentrated in the United States; and in August of that year the programme was code-named 'The Manhattan Project'.

Although the Manhattan Project eventually proved remarkably successful, initially many blunders and mistakes were made and many blind alleys taken. The central problem was to develop methods that could produce sufficient quantities of fissile material – either relatively pure U-235 or Pu-239 – for one or more nuclear weapons.

In ordinary uranium, atoms of U-235 occur in the proportion of only one atom in 140; the remaining 139 are atoms of U-238. To obtain relatively pure U-235

The Hanford plutonium production facility – a part of the Manhattan project – on the banks of the Columbia River whose waters cooled the nuclear reactors. The urgent construction of this enormous plant during the war necessitated the employment of 42,400 workers and the development of new skills and techniques

21

it is therefore necessary to separate it from U-238 in uranium. By 1942, research on three methods of U-235 separation was far enough advanced to justify an attempt at large-scale production – the gaseous diffusion, electromagnetic, and centrifuge methods.

The production of Pu-239 from uranium in appreciable amounts was connected with the operation of a controlled chain reaction. Natural uranium by itself cannot be used to produce a chain reaction because of the large proportion of U-238 contained in it. But in 1940, Fermi and Szilard, then working in the United States, suggested a way round this difficulty. This was to mix natural uranium with a substance whose nuclei are small in size so that if a fast neutron collides with one of them it will lose a large fraction of its velocity – just as a billiard ball will lose velocity when it collides with another one. The neutron's velocity is thus rapidly 'moderated' down to the low velocity at which it can be efficiently captured by a U-235 nucleus, producing fission, and at which it will have a relatively high probability of avoiding capture by a U-238 nucleus.

A substance used to slow down fission neutrons is called a moderator. And a good moderator reduces the velocity of fast neutrons to low values in a small number of collisions without the nuclei of the atoms of the moderator capturing the neutrons to any great extent. Fermi and Szilard suggested the choice of carbon, in the form of graphite, as a moderator because of its availability at that time.

The minimum condition for maintaining a chain reaction is that, for each nucleus undergoing fission, at least one fission neutron causes the fission of another nucleus. This condition is expressed in terms of a multiplication factor, defined as the ratio of the number of neutrons of any one generation of fission to the number of neutrons of the preceding generation. The multiplication factor in a uranium-moderator system will be determined, on the one hand, by the number of neutrons lost in the system by non-fission capture by U-238, by non-fission capture by nuclei of the moderator and other substances (such as impurities in the moderator, and the uranium and its fission products), and by escape through the exterior surface and, on the other hand, by the number of neutrons generated by the fission of U-235 by slow neutrons. And if the number of neutrons produced in the latter process exceeds the total number lost by the first three processes there will be a net gain; the multiplication factor will exceed one and a chain reaction will be possible. The size of the system in which the number of neutrons lost just balances the number produced by fission is called the critical size.

Towards the end of 1942, a self-sustaining chain-reacting system, using natural uranium and graphite, was constructed on a squash court at the University of Chicago. A major obstacle was the production of the large amounts of pure uranium metal required – the total amount of uranium produced in America up to

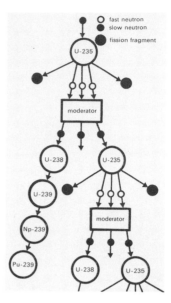

The chain reaction as in the nuclear reactor showing (above) the fission of uranium-235 and the subsequent capture of neutrons by uranium-238 leading to the production of plutonium, an element not found in nature

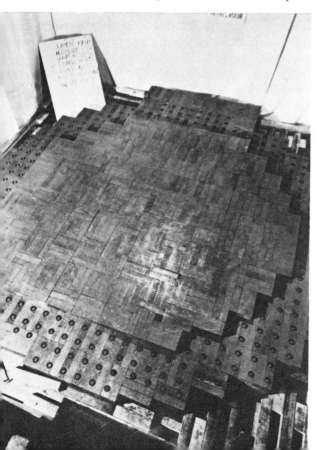

Left: the only known photograph of the world's first nuclear reactor, the Chicago 'pile', taken during the addition of the 19th layer of graphite

this time was only a few grams and even this was of doubtful purity! But the production of pure graphite proved to be less difficult, even though the degree of purity required was hitherto unheard of. The system consisted of lumps of uranium, separated by a distance of about 20cm with graphite bricks. About 40,000 graphite bricks were needed, each one about 10cm wide and deep and about 42cm long. A layer of solid graphite bricks alternated with a layer of graphite bricks in which holes were bored to take pieces of uranium. Only about 6 tons of uranium metal were available at the time and, as this was insufficient, some uranium oxide had to be used. Because the structure was made by piling one layer on top of another it was called a pile. It was to be approximately spherical in shape, about 8 metres in diameter.

Cadmium strips were inserted into the pile during its construction as a safety measure. Cadmium has the property of absorbing neutrons very efficiently, and the strips proved to be a necessary precaution since the critical size was reached sooner than was anticipated. For this reason, the finished pile was not spherical as planned but was flattened at the top.

By withdrawing the cadmium strips at intervals and measuring the neutron intensity inside the pile, the approach to critical size was observed. At 3.25 p.m. on 2 December 1942, a sharp increase in neutron intensity occurred, which showed that a self-sustaining chain reaction had been initiated for the first time.

The realization of a chain reaction in a uranium-graphite system opened up the prospect of the production of appreciable quantities of plutonium. The neutrons absorbed by U-238 without producing fission would then not be wasted since the U-239 would be transformed into fissile Pu-239.

But the Chicago pile itself was not suitable for plutonium production, because it would have been necessary to dismantle it to get at the plutonium. In any case, the pile had no cooling system, and therefore the safe operating level was so low that only traces of plutonium were produced.

The complete fission of a single gram of U-235 per day produces about 1 megawatt of power – enough to run 1,000 cooker plates – most of which appears as heat. If one neutron per fission is captured by a U-238 nucleus to form Pu-239, then the fission of 1 gram of U-235 per day would produce 1 gram of Pu-239 per day. Thus, a pile (or a reactor as it is now called) constructed for the production of 1 kilogram of plutonium per day would produce energy, mostly as heat, at a rate of 1,000 megawatts. This is an enormous quantity of heat and the problem of cooling such a reactor is considerable.

In 1942, none of the four rival processes for the large-scale production of fissile materials, namely the three methods of producing U-235 and the reactor production of plutonium, offered the certainty of success. Moreover, each one was extremely complex and expensive, and it could not be predicted which would produce adequate fissile material in the shortest time. Speed was of the utmost importance since it was thought that the German programme was probably farther advanced, and it was at first decided to try all four methods simultaneously. But the centrifuge was soon discarded and efforts concentrated on gaseous diffusion and the electromagnetic process for U-235 production, and the uranium-graphite reactor for plutonium production. Large plants on a scale never attempted before had to be built for each process, new industries created, equipment and technical processes developed from scratch, and all this in the shortest possible time – a colossal scientific and technological undertaking, probably only possible under war-time conditions and, even then, only because of the incalculable consequences of the likely results.

U-235 and U-238 are chemically identical and so it is necessary to use a physical method to separate them. But plutonium can be separated from uranium by a chemical procedure because the atomic numbers of these two elements differ by two units. For use in nuclear weapons, the concentration of U-235 in ordinary uranium has to be increased from its 'natural' value of 0.72 per cent in uranium ore to a value of over 95 per cent, a process normally called enrichment. The gaseous diffusion method of enrichment is based upon the fact that, in a gaseous mixture of two isotopic molecules, the molecules of the lighter isotope will diffuse more rapidly through a porous barrier than those of the heavier one.

But if gas diffusion was to work, it was first necessary to choose a uranium compound which could be readily turned into a gas. Uranium hexafluoride UF_6 which is solid at room temperatures but easily vaporized, was the best choice. But because UF_6 gas is extremely corrosive and reactive, special materials had to be used for the construction of the pipes and pumps used in the process, and the entire installation had to be completely free from grease and oil. Furthermore, the fact that the proportion of U-235 is raised by only a small factor in each diffusion stage means that a large number of stages were required to obtain the desired enrichment. For example, about 4,000 stages would be required for an enrichment of 99 per cent.

The gaseous diffusion plant built as part of the Manhattan Project was of a gigantic size, by far the largest industrial undertaking ever attempted. Construction began at Oak Ridge, Tennessee, in mid-1943. The plant covered nearly 50 acres of floor area and contained hundreds of acres of diffusion barriers connected by an incredible maze of hundreds of miles of piping with thousands of valves and joints, all requiring intricate welding. The precision with which the components of the plant had to function was unheard of in engineering history. And the process was fully automated – the first ever large-scale automation. The plant consumed enormous quantities of electrical power, requiring the construction of an independent power station – the largest one existing at that time. The enterprise was indeed a monumental undertaking. The plant is still used for the production of enriched uranium, although the output is now mostly employed for peaceful purposes.

But the first method actually to yield appreciable

quantities of U–235 for the Manhattan Project was the electromagnetic method, which is based on the principle that each species of isotope present in a stream of ions is bent through a different path by electric and magnetic fields. The isotopes can then be collected separately. In September 1942, the construction of an electromagnetic-separation plant, consisting of a number of independent units, was begun at Oak Ridge. Once again, this was an incredible undertaking. The method had not even been completely developed and the magnets, electric power supplies and other equipment were of radically new design. The magnets were huge structures, 30 metres in length and each containing thousands of tons of steel. And a vacuum system had to be installed capable of evacuating huge volumes to a very high vacuum. Again, a system of this magnitude had never before been attempted.

The efficiency of the electromagnetic method was found to increase dramatically when slightly enriched uranium rather than ordinary uranium was used as the input material. Consequently, it was decided to use a thermal-diffusion technique to produce feed material for the electromagnetic plant – a comparatively simple method consisting of two long, vertical, concentric pipes enclosed in a cylinder. The inner pipe was steam heated from the inside and the outer one water cooled from the outside. When liquid UF_6 under pressure was circulated between the pipes, molecules of the lighter isotope tended to concentrate near the hotter surface. But the technique was suitable only for the production of slightly enriched material since enormous amounts of steam had to be used and the quantity of coal which would have been required for high levels of enrichment was simply not practicable.

A thermal-diffusion complex, consisting of about 2,000 columns, each nearly 50 feet high, was constructed close to the power station which had been built to supply the gas-diffusion plant. This plant had not been completed at this time and, therefore, the steam required for thermal diffusion could conveniently be produced by the power station.

The Oak Ridge establishment finally consisted of three huge complexes – one housing the gas-diffusion plant, another the electromagnetic plant and a third the thermal-diffusion set-up. It was decided to postpone completion of the final stages of the gas-diffusion system and to use the less highly enriched output from the earlier stages as input material for the electromagnetic plant. And so the final product came from the electromagnetic plant, fed from both the gas-diffusion plant and the thermal-diffusion plant. At that time, this was the fastest way of producing the required uranium. The gas-diffusion plant started feeding the electromagnetic plant with 7 per cent enriched uranium in June 1945. And enough weapons-grade U-235 was produced within one month to make one nuclear bomb.

The plutonium required for the Manhattan Project was produced by uranium-graphite reactors, one built at Oak Ridge and three others at Hanford, Washington. The Oak Ridge reactor was air cooled but the much larger Hanford reactors were water cooled. The smaller one, which started operating in November 1943, was built mainly to provide badly needed plutonium for experimental purposes, particularly for testing possible methods for the chemical separation of plutonium from uranium and its fission products. The reactor consisted of a cube of graphite containing a number of horizontal channels. And fuel, in the form of metallic uranium cylindrical elements encased in aluminium, was slid into the channels in the graphite. When the uranium elements were ready to be sent for separation they were pushed through the back of the reactor into tanks of water. There they were left for a few days to allow time for the U-239 to change into Pu-239. They were then dissolved in acid and the plutonium was chemically separated from the uranium and the fission products. But the rate of plutonium production from the Oak Ridge reactor was small, only a few grams per month.

The construction of the first large reactor at Hanford was started in June 1943 and operation began in September 1944. The second and third reactors were completed a few months later. The construction of these reactors and their separation plants was yet another spectacular achievement. There had been precious little previous experience with chain-reaction systems. And it is remarkable that the separation plants, which were operated entirely by remote control because of the danger of radioactive contamination, were based on initial studies made with only half a milligram of plutonium! The production of appreciable quantities of plutonium began at Hanford in February 1945.

Fat Man and Little Boy

By July 1945, enough plutonium had been produced to construct two bombs but there was only enough U-235 for one bomb. It was, therefore, decided to test just one (plutonium) bomb and to hold the other two in reserve in case they were needed as weapons at some future date.

In the uranium bomb, which became known as 'Little Boy', a sub-critical mass of U-235 was fired down a 'cannon barrel' into another sub-critical mass of U-235 placed in front of the 'muzzle'. And when the two masses contacted they formed a critical mass which exploded. About 15 kilograms of U-235 were required and a special gun had to be designed for this purpose – light and short enough to fit into an airborne weapon but sufficiently fast-firing to ensure that the critical mass was formed in a very short time. Otherwise the bomb would pre-detonate because neutrons emitted by one of the uranium pieces would begin interacting with the other piece before the mass became critical.

But the gun-assembly method could not be used for the plutonium bomb, known as 'Fat Man'. The characteristics of the fission of plutonium make the method too slow to prevent a pre-detonation. Instead, an implosion technique was used. A plutonium core, in the form of two gold-clad hemispheres, was surrounded by a tamper made of platinum. The costly

*The two nuclear bombs of approx-
imately 20-kiloton capacity used in
1945. 'Fat Man' (background) is
the implosion-type plutonium device
detonated over Nagasaki, and 'Little
Boy' the gun-type, using fissionable
uranium, dropped on Hiroshima.
Compared to modern nuclear weapons these devices were very primitive*

tamper had two purposes. It would reflect back into
the plutonium some of the neutrons which escaped
through the surface of the core, allowing some
reduction in the mass of plutonium needed. But a more
important function was that, because the tamper was
made of heavy material, its inertia would hold together
the plutonium during the explosion for a fraction of a
second and thus contain the disintegration of the
exploding bomb to obtain greater efficiency. The
plutonium core was surrounded by TNT, arranged as
explosive lenses focused on to the plutonium sphere.
These were to be detonated perfectly simultaneously
so that the entire surface of the sphere was compressed
with perfect uniformity. The mass of plutonium (approximately 5 kilograms) in the core was just less than
the critical mass, but the compression was designed to
compress the core into a critical mass by increasing its
density.

The final component was the so-called 'Urchin',
required to initiate the fission reaction in the plutonium
at the moment of implosion. The Urchin consisted of
a small hollow sphere placed at the centre of the plutonium core. Inside was some beryllium and polonium,

two elements which produce neutrons when intimately mixed. Explosive lenses were focused onto the surface of the Urchin. The two substances were placed separately on the inside of the Urchin so that no neutrons were produced until the moment of implosion. And then the Urchin would be crushed, the beryllium and polonium mixed, and a pulse of neutrons given off at precisely the right instant to initiate the nuclear explosion. The timing of the detonation of the explosive lenses was crucial for the operation of the implosion bomb. Microsecond precision was essential.

The first nuclear weapon test, given the code name 'Trinity', was performed on 16 July 1945 in the Alamogordo desert, New Mexico, also aptly named in Spanish 'Jornado Del Muerto', – 'Journey of Death'. It is about 40 miles across and supports little life apart from cacti and some insects. The bomb, which weighed five tons, was exploded at the top of a hundred-foot tower.

After observing the explosion from the observation post about two miles away from the tower, Fermi wrote: 'My first impression of the explosion was the very intense flash of light, and a sensation of heat on the parts of my body that were exposed. Although I did not look directly towards the object, I had the impression that suddenly the countryside became brighter than in full daylight; I subsequently looked in the direction of the explosion through the dark glasses and could see something that looked like a conglomeration of flames that promptly started rising. After a few seconds the rising flames lost their brightness and appeared as a huge pillar of smoke with an expanded head like a gigantic mushroom that rose rapidly beyond the clouds, probably to a height of the order of 30,000 feet. After reaching its full height the smoke stayed stationary for a while before the wind started dispersing it.' It was natural that the scientists observing the test felt proud and relieved that the long and hard effort had been finally successful. But many of them were later to have second thoughts about their fateful creation.

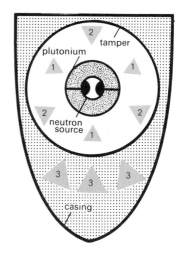

The implosion method of detonating plutonium nuclear weapons. The chemical explosive lenses marked 1 are focussed on the 'Urchin', 2 onto the surface of the plutonium sphere and 3 onto the tamper

Little Boy and Fat Man were dropped from aircraft over Hiroshima and Nagasaki on 6 and 9 August 1945 respectively, each bomb releasing roughly the same quantity of energy as 20,000 tons of TNT. The *complete* fission of one kilogram of U-235 or Pu-239 would release about this much energy. Therefore, the uranium bomb had an efficiency of about 10 per cent and the plutonium bomb a slightly higher value – about 20 per cent.

Fusion bomb

Since 1945, several 'improved' types of nuclear weapons have been developed. The most important of these is the thermonuclear weapon which uses the energy produced by a type of nuclear reaction called nuclear fusion. This process involves nuclei of very low atomic number and is, therefore, at the opposite extreme to fission which involves nuclei of very high atomic numbers. Fusion occurs when a pair of light nuclei unite together to form a nucleus of a heavier atom. For example, two deuterium nuclei can, under suitable conditions, combine to form a nucleus of the element helium. Deuterium is an isotope of hydrogen often called heavy hydrogen. Whereas the nucleus of an atom of ordinary hydrogen contains just a proton, a deuterium nucleus contains a proton and a neutron. Tritium, a third isotope of hydrogen, has also been used to produce fusion.

The fusion process in which two deuterium nuclei fuse to produce helium 3, and a fast neutron giving rise to a large amount of energy

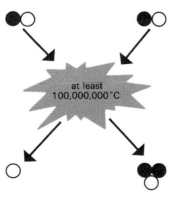

at least
100,000,000°C

Fusion is accompanied by a substantial release of energy – equal to the energy equivalence of the difference between the total mass of the fusion products and the total mass of the two nuclei which fuse together. The fusion of all the nuclei present in 1 kilogram of deuterium would release roughly the same amount of energy as the explosion of 60,000 tons of TNT, or the burning of 10,000 tons of coal. Nuclear fusion is brought about by means of a very high temperature – hence the term thermonuclear.

During the fusion of, for example, deuterium, high-energy neutrons are emitted which can be used to cause fission in both U-235 and U-238. And, if natural

uranium is used in association with fusion, a large
amount of additional energy can be obtained from
fission. A temperature of several millions of degrees is
necessary to make fusion reactions take place – in a
thermonuclear weapon this is obtained by a fission
explosion. But, for technical reasons, it is preferable to

*This radioactive dust from a nuclear
explosion in the Nevada Desert will
encircle the earth in the upper
atmosphere and be precipitated as
radioactive 'fallout'*

use U-235 rather than Pu-239 in the fission device. The thermonuclear explosion is obtained by combining a quantity of deuterium or tritium or both with the trigger. To make use of the thermonuclear neutrons, the fusion weapon can be surrounded by a blanket of ordinary uranium. The resultant explosion is then a combination of the fission energy from the triggering device, thermonuclear energy and additional fission energy produced by U-235 and U-238 in the blanket. A fission-fusion-fission weapon can produce tremendous explosive power, equivalent to tens of millions of tons of TNT; this is three orders of magnitude (1,000 times) larger than the yield of the first nuclear weapon.

The possibility of developing a thermonuclear weapon was explored by American scientists as early as 1942, and feasibility studies were conducted as part of the Manhattan Project. But they were strictly subordinated to the development of the nuclear weapon because it was believed that this could be obtained more quickly. After the end of the Second World War a small research programme on thermonuclear energy was continued. Then, on 23 September 1949, the Soviets unexpectedly exploded their first nuclear weapon and, as a result, attention in the United States was again directed to the question of developing a thermonuclear weapon. Finally, on 31 January 1950, President Harry S. Truman issued an order to the Atomic Energy Commission to proceed rapidly with the development, even though a majority of the Commission advised against it.

The first thermonuclear weapon was exploded by the United States on 1 November 1952; the Soviet Union exploded one less than a year later, on 21 August 1953.

Three more nuclear weapon states have since emerged – the United Kingdom, France and China – each of which has acquired both nuclear and thermonuclear weapons. And the five powers have, between them, tested many hundreds of these weapons. The actual number may well exceed 1,000 – equivalent in explosive power to a total of over 500,000 tons of TNT

– about 100 times the total of all the explosives used in war since gunpowder was invented.

President Truman, in his October 1945 message to Congress, spoke of nuclear energy in the following terms: 'Never in history has society been confronted with a power so full of potential danger and at the same time so full of promise for the future of mankind and the peace of the world. I think I express the faith of the American people when I say we can use the knowledge we have won, not for devastation of the world but for the future welfare of mankind.' In this instance, faith did not move mountains – the fulfilment of the promise of peaceful nuclear energy was to be long delayed.

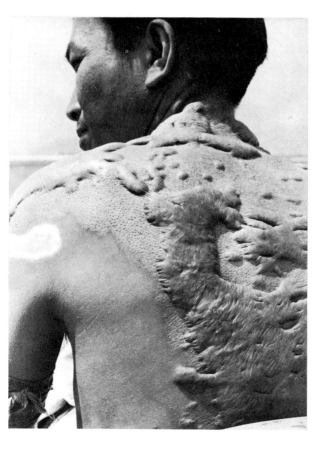

Having survived the explosion with severe burns this Hiroshima casualty may still suffer from malignant cancer or leukaemia as a result of exposure to radiation. He might also be infertile or his offspring may be adversely affected by genetic mutations

35

ATOMS FOR PEACE

2

On 8 December 1953 President Eisenhower, in his famous 'Atoms for Peace' speech to the United Nations in New York, said: 'The United States knows that peaceful power from atomic energy is no dream of the future. That capability, already proved, is here – now – today. Who can doubt, if the entire body of the world's scientists and engineers had adequate amounts of fissionable material with which to test and develop their ideas, that this capability would rapidly be transformed into universal, efficient, and economic usage?' About one year later, the first nuclear power reactor to supply useful amounts of electricity started up in the Soviet Union.

The ten-year delay in the development of peaceful nuclear power was caused by the advent of the cold war which so closely followed on the heels of the Second World War. The high level of international tension had ensured that nuclear energy resources continued to be mainly concentrated on military applications.

In 1946, the United States Congress passed the MacMahon Act, and thereby adopted a policy of 'nuclear isolationism'. And, internally, the development of nuclear energy remained under Federal control, firmly in the hands of the Atomic Energy Commission. But, under mounting pressure from private industry, the AEC began to examine seriously the possibility of designing reactors able to generate

This 100-foot-high aluminium-sheathed building at Garching, Germany, houses a 1,000 kilowatt 'swimming-pool' reactor. Many countries use this type of research reactor to gain experience in nuclear technology, to train personnel in nuclear operations and for radio-nuclide production

electricity as well as to produce weapons-grade plutonium. By 1951, the Commission had begun to work in close harmony with the public authorities responsible for electricity production. Up to this date, the paltry non-military efforts were mainly concentrated on the development of reactors which might power ships and aircraft. But, in the early 1950s, the Americans became convinced that nuclear power could be rapidly developed as a source of cheap energy, particularly for the developing countries – as was clearly evident in the 'Atoms for Peace' plan. Subsequent events have shown that American confidence in the imminence of low-cost nuclear energy, based as it was on very limited experience, was more than a little premature.

Reactors and research

During the decade 1945-54 four new countries, in addition to the United States and the Soviet Union, initiated nuclear energy programmes – Canada, France, Norway and the United Kingdom. Until the latter part of the period, the British and French programmes were geared to the production of plutonium for nuclear weapons whereas efforts in Canada and Norway were entirely put into peaceful applications. But there was one common factor – all four countries constructed research reactors.

Research reactors are used essentially as sources of neutrons: for the production of artificial radionuclides; for studying the effects of neutron bombardment on materials selected for the construction of power reactors; for gaining experience in reactor operation and in training personnel; and for pure research, mainly to establish the parameters and check the theory of reactor physics, but also in chemistry, biology, medicine and physics.

By 1954, there were 28 research reactors in operation: 15 in the United States; 5 in the Soviet Union; 4 in the United Kingdom; 2 in Canada; 1 in France; and 1 in Norway. And, by 1969, 375 were operating in 50 countries, 41 of them in developing countries.

There are several types of research reactor of which the most widely used and versatile is the pool reactor. Many universities throughout the world now have one. This type of reactor is built in a large pool resembling a swimming pool at least 18 feet deep, but with no shallow end. This depth is required to provide sufficient shielding to reduce the radiation from the core to safe levels. Water in the pool acts as the coolant, and often as the moderator, as well as the shield. The core of the reactor, containing the uranium fuel usually in the form of rods encased in stainless steel, is suspended from a bridge which runs across the pool. The bridge is provided with wheels so that it can be moved along the sides of the pool. One of the main advantages of the reactor is that it can be placed in any part of the pool. An experiment using the reactor may be proceeding in one part while a second experiment is being set up or dismantled in another.

The most important use of research reactors is for the production of radionuclides. If a nucleus captures a neutron, and virtually all nuclei will do so, the new nucleus formed is usually radioactive. A neutron is most likely to be captured if it is moving slowly, rather in the way that a slow golf ball will drop into the hole whereas a fast one will jump over. Radionuclides are normally produced by bombarding stable (non-radioactive) elements with slow neutrons inside the core of a research reactor. For example, to produce the radionuclide sodium-24, which is much used in medical and industrial applications, a can containing ordinary sodium (Na-23) is placed in one of a number of compartments specially provided in the core of the reactor. After a suitable time, the can is removed and much of the stable sodium will have been converted into the radioactive isotope by the neutron bombardment. In Great Britain, many species of radionuclides are produced in various reactors and marketed by the Radiochemical Centre at Amersham. Sales now exceed £2.5 million annually from nearly 60,000 shipments of radionuclides, mostly overseas. Some of the uses of these isotopes will be discussed in chapter 4.

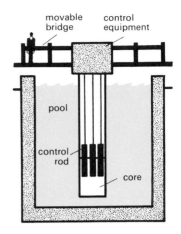

The swimming-pool type research reactor

The BEPO reactor which in its day was a major producer of radioactive isotopes for medical and industrial uses. Above: it is loaded with materials to be irradiated at high temperatures – work which was useful in designing full scale nuclear power reactors. Above, right: the full view

An historically famous research reactor was the British Experimental Pile Operation, or BEPO, which started operating in July 1948 and which has only recently been shut down. BEPO was built in a disused aircraft hangar at the Atomic Energy Research Establishment, Harwell, and was, for several years, one of the world's major producers of radionuclides. It also originated a family of reactors which is still growing.

The moderator was carbon, in the form of graphite, and the fuel was rods of natural uranium canned in aluminium. The core consisted of a cube built of graphite blocks arranged so as to leave horizontal channels for the fuel rods. BEPO was controlled by pushing steel rods loaded with boron – a neutron-absorbing material – into another channel in the core. To shield the workers at the reactor from the harmful effects of radiation passing out of it, the core was built inside a concrete box, thick enough to reduce the radiation to a safe level.

BEPO was cooled by air, drawn through the channels by powerful fans. After filtration to remove any radioactive dust the hot air was discharged up a 200-foot chimney. In 1951, a heat exchanger was installed in the air duct to supply hot water for the heating system of neighbouring laboratories – the very first peaceful application of fission energy.

The first two British plutonium-production reactors were built at Windscale, on the Cumberland coast, on essentially the same pattern as BEPO. But, because they operated at much higher power, the cooling system presented novel problems. It was originally intended that the reactors should be cooled by water like the Hanford ones, but it proved difficult to find a suitable site in the highly-populated British Isles for a water-cooled reactor. The main factor to be considered was safety since, if some of the water accidentally escaped from the cooling circuit, a number of the neutrons, which would otherwise have been captured, would become available to cause fissions. The reactor might then become uncontrollably overheated. Furthermore, water has the disadvantage that its hydrogen captures neutrons appreciably and this would seriously reduce the efficiency of the system. For these reasons, it was decided to use a gas as the cooling fluid.

Gases are much less efficient at removing heat from surfaces than liquids and to adequately cool a cylindrical fuel element the gas would have to flow past it at a fantastic rate. However, a simple solution to this

An aerial view of the British nuclear complex of Calder Hall and Windscale, which, although built initially for the nuclear weapons programme, is now a vital element of the British peaceful nuclear programme. The Windscale experimental advanced gas-cooled reactor is to the right of the Calder Hall reactors

Plutonium fuel elements manufactured at Windscale for experimental use in the breeder-reactor at Dounreay. Developed to a universal design, British power reactors will produce enough plutonium to fuel many breeder reactors

difficulty was found – the cans encasing the fuel elements were finned. This trick increased the surface area in contact with the gas and also increased turbulence. Consequently, heat was transferred with a very much greater efficiency. In true British tradition, the cheapest gas – air – was used in the simplest manner – at atmospheric pressure.

As in BEPO, the natural uranium fuel elements were placed in horizontal channels in graphite. Cooling air was blown over the elements by very powerful fans, and then discharged through a chimney after filtration. When the fuel elements had remained in place for a long enough time to build up sufficient plutonium, they were pushed through the back of the reactors into a tank of water, left for a few weeks to allow the fission products to decay and then sent to a reprocessing plant for extraction of the plutonium.

In 1957, a miscalculation during a routine operation involving the fuel elements in one of the two Windscale reactors caused some uranium fuel to become overheated. Fire broke out and the combustion products – including some radioactive material – escaped into the atmosphere. When teams of scientists began to test for radioactive contamination in the neighbourhood they found traces of radioactive iodine in the local milk. Milk deliveries were then banned in an area of 200 square miles. A good deal of alarm was created up and down the country by the incident – most of it unnecessary because, so far as is known, no injury or damage was in fact sustained. The Atomic Energy Authority, however, promptly issued regulations designed to prevent a recurrence of the mishap. After this accident both reactors were shut down, but during their operating life they yielded, in addition to plutonium, much experience in reactor operation which was used to good effect in the design of larger versions to be built at Calder Hall, primarily to produce plutonium but also to generate electricity.

The French programme closely paralleled that of the British. An Atomic Energy Commissariat was set up in 1945 and, seven years later, it initiated a five-year programme which entailed a complex of three reactors at Marcoule. One of these, the G-1, was completed in 1956 and was built for the production of plutonium for nuclear weapons. But G-2 and G-3 are dual-purpose reactors, producing electric power as well as plutonium. The G-1, like its Windscale counterpart, is air cooled, graphite moderated and uses natural uranium fuel.

In view of the experience gained from their military and research reactors, it is not surprising that the early nuclear-weapon powers were the first to generate useful amounts of electricity from experimental power reactors. In 1954, a small, light-water moderated, gas-cooled reactor started up at Obninsk in the Soviet Union, generating 5 megawatts of electricity (MWe). Two years later, the first of four 40-MWe gas-cooled reactors went into operation at Calder Hall. In 1957, a 60-MWe light-water reactor began producing power at Shippingport in the United States. And in 1958, the

The hall in the French nuclear establishment at Marcoule in which plutonium is extracted from spent reactor elements. The French, unlike the British, are now favouring light-water reactors rather than gas-cooled.

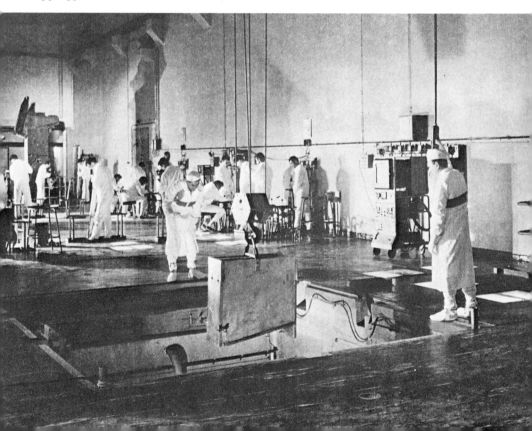

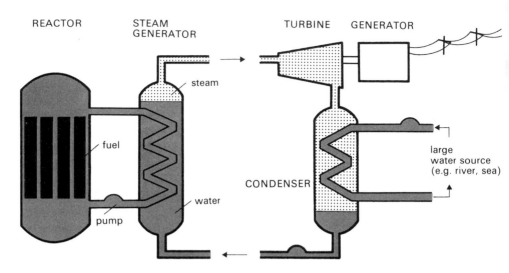

REACTOR   STEAM
          GENERATOR   TURBINE   GENERATOR

steam

fuel

large
water source
(e.g. river, sea)

CONDENSER

water

pump

The main components of a nuclear power station. Heat is removed from the reactor to produce steam which, in turn, is used to drive the generator turbine before being condensed for recirculation. For economic reasons, most modern nuclear power stations produce at least 500 megawatts of electricity

French 40-MWe G-2 gas-cooled reactor started up at Marcoule. The first power reactor to operate outside these four countries was the West German 15-MWe light-water one at Grosswelzheim, in November 1960.

A power reactor is basically a furnace where a self-sustaining chain reaction can be controlled and the heat put to useful work. In a typical case, the fuel, moderator, control rods, and coolant are enclosed in a pressure vessel. The coolant, heated by its passage over the fuel elements, flows through a heat exchanger where it turns water in a secondary circuit into steam, used to drive a turbine generator which produces electricity.

The radioactivity produced by fission is confined to the reactor core. But, immediately surrounding the core, radiation levels are extremely high and it is essential to shield workers in the vicinity. The massive shields, placed inside or outside the pressure vessel, are made from dense materials like concrete or steel.

Like all power producers, a reactor must be provided with a system of controls to allow normal starting and stopping, and rapid shut-down in case of an emergency. Control is usually achieved by rods, containing neutron-absorbing material such as boron or cadmium, which are moved automatically in and out of the core to govern the rate at which fission takes place. In

In an expensive and dramatic experiment in the interests of formulating safety procedures a boiling-water reactor was deliberately allowed to 'run away'. The control rod was suddenly removed and within $\frac{1}{10}$ second the power output was increased to 10 million kilowatts. In the resulting explosion machinery was hurled 80 feet into the air

45

practice, more than the minimum amount of fuel is initially fed into the core and the control rods are used to mop up the excess neutrons. The rods are withdrawn as the fuel burns away so that the multiplication factor is maintained at a value of one. But there is little chance of a reactor exploding since, if the rate of fission ever became too great and excessive heat was generated, the system would break up and the chain reaction cease. Melting is more likely than an explosion.

There are three main types of power reactors, each characterized by the material used as the moderator: graphite, light-water (ordinary), and heavy-water (in which the hydrogen is replaced by deuterium).

Graphite-moderated reactors

Graphite-moderated, gas-cooled reactors have received most attention in Great Britain and France, mainly because they were first adopted for plutonium production. It was believed that they could be rapidly developed to a stage at which they would produce power economically, using conventional techniques and cheap materials. In Britain, the decision was made in 1950 that plutonium, additional to that from the Windscale reactors, should be produced by units also capable of producing power. And so, when the Chiefs of Staff pressed for an increase in the production of military plutonium, it was decided that it should come from a dual-purpose plant. This was built near the Windscale plutonium-separation factory on land which had formerly belonged to Calder Hall, a seventeenth-century mansion. Access to the remote site is by a road over a specially-built bridge spanning the River Calder. No railway is needed since a year's fuel supply can be delivered by a few lorries compared with the thousands of railway wagons that are required to transport the fuel for a coal-fired station. Nor is the countryside marred by ashtips. The spent fuel elements are simply taken by road to the Windscale factory for reprocessing.

Calder Hall consists of four reactors in which the heat generated is not wastefully thrown away but is used instead to produce electricity. And, since the efficiency with which heat can be converted to electricity is proportional to the operating temperature, the reactors had to be run at considerably higher temperatures than those at Windscale. Attention had again to be directed to the problem of cooling. Because graphite was the obvious choice of moderator, the use of air for cooling was out of the question since the rate of oxidation of the graphite at such high temperatures would be too great. A more suitable gas had to be chosen and, of the many considered, carbon dioxide was finally picked, on account of its cheapness, availability and good heat-transfer properties. But, to obtain a high cooling efficiency, it had to be used at a pressure well above atmospheric, which meant that the reactor cores had to be contained in strong pressure vessels.

Moreover, a new material, stable in an atmosphere of carbon dioxide at the high operating temperatures, was needed to can the fuel elements – aluminium would certainly not do. For this purpose, materials technologists developed a new alloy of magnesium and beryllium, called Magnox.

Each of the four reactors, with its shield and associated buildings, weighs over 20,000 tons and rests on a reinforced concrete raft, 130 feet by 104 feet and 11 feet thick, which weighs another 10,000 tons.

Very high quality reinforced concrete was needed for the raft so that it would not move when completed otherwise the shield might fracture. The 7 foot thick concrete shield is octagonal, about 88 feet high and 46 feet across inside. The internal walls had to be built to within a quarter of an inch of the vertical throughout their height – a tough specification for civil engineers. The cylindrical pressure vessel was built in 90-ton sections which were lifted into the octagon by a 100-ton derrick on a 90-foot steel tower, nicknamed 'the big stick'. The vessel is constructed from 2-inch thick manganese steel and is 37 feet across and 71 feet tall.

The most complex operation of all was the construction of the reactor core inside the pressure vessel. The

Herringbone type fuel elements being loaded into a fuel basket for the reactors at Calder Hall. The development of Magnox, which sheathes the element, was one of the triumphs of materials technology which has contributed to modern power production

core is also cylindrical, 27 feet high and 35 feet across, and is made from 58,000 interlocked graphite blocks. Around the outside of the cylinder are 11 restraint rings to hold the structure together. The core is pierced by 1,696 vertical fuel element channels and is built in such a way that, no matter what dimensional changes occur during the lifetime of the reactor, the size and spacing of the channels will not be disturbed.

The 10,176 fuel elements are made of natural uranium rods, 1.15 inches in diameter and 40 inches long. These Magnox-clad elements are finned in a helical fashion rather than along their lengths as in the Windscale reactors. Six elements occupy each channel and the total weight of uranium in the reactor is about 120 tons. Fuel elements are put into and taken out of the core through a movable chute which enters the pressure vessel through a tube at the top and is poked into the appropriate channel. An electrically-controlled grab is lowered down the chute at the end of a cable and takes hold of the element. The cable is then wound in, bringing the element with it.

The carbon dioxide enters the core at a pressure of seven atmospheres and a temperature of $145°C$, and leaves it at a temperature of $340°C$. The flow rate is about 7 million pounds per hour and the flow velocity is 42 feet per second. The hot gas is taken from the pressure vessel to heat exchangers, of which there are four per reactor, through steel ducts $4\frac{1}{2}$ feet in diameter. The cool gas leaving the exchangers is returned to the reactor through similar ducts in which blowers are placed to circulate the gas. The steam from the heat exchangers is fed, at a temperature of $315°C$ and a pressure of 210 pounds per square inch, to two turbo-electric generators.

The power output of the reactor is controlled by moving control rods in and out of channels in the core. There are 48 rods, made of 4 per cent boron-steel and clad in stainless steel, suspended from steel cables wound on a drum.

The first objects to be seen when approaching Calder Hall are four 300-feet high towers. These are diabolo-

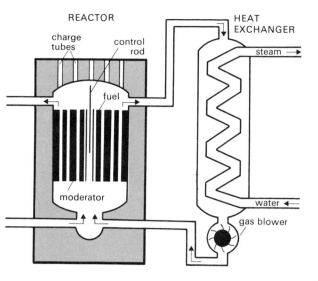

REACTOR

HEAT EXCHANGER

charge tubes

control rod

steam →

fuel

moderator

water ◄

gas blower

The Calder Hall gas-cooled reactor. The coolant in all AGRs is carbon dioxide but this will probably be superseded by helium

shaped concrete cooling towers built over shallow ponds, and their purpose is as follows. When the steam from the heat exchangers has passed through the turbines, it is condensed by flowing it over tubes, in condensers, through which cold water runs. This water is warmed during the process and is cooled in the towers so that it can be re-used. Hundreds of millions of gallons of water a day flow around the plant and it is, therefore, essential that as much of this as possible should be conserved. The warm water from the condensers is led into a tower at points near its base and is allowed to fall in a spray into the pond. The water is cooled as it falls by a current of air which rises up the tower by convection. Even so, a few million gallons of the water used in the condensers are lost by evaporation each day and this has to be replaced from local rivers.

The first reactor at Calder Hall began operating on 22 May 1956, and the second in December 1956. The successful commissioning of the world's first large-scale nuclear power station was an historic occasion. And, at the opening ceremony, H.M. the Queen said that the reactors 'offer us a vital and timely addition to the industrial resources of our nation and to our material welfare. But above all we have something new to offer to the peoples of the underdeveloped and less fortunate areas of the world who will continue to look to us for assistance and example as they have done in the past'. Once again, these hopes were to be proved over-optimistic.

The two remaining reactors started up in March and December 1958. Calder Hall feeds nearly 150 MWe into the national grid, enough to supply the electricity needs of a city of about a million inhabitants. And four more reactors, essentially identical to those at Calder Hall, were built at Chapel Cross in Annan, Scotland. They came into operation between November 1958 and December 1959.

The 500 MWe Magnox power stations at Hinkley Point on the north coast of Somerset. The tanks in the foreground are for carbon dioxide storage. Operating experience from such stations has overcome the initial teething problems so that efficiency comparable to that of conventional stations is now obtained

The successful operation of these eight prototype reactors, which have been compared with the Model-T Ford car, led to the construction of larger reactors designed to optimize the production of power rather than plutonium. The first two of these all-power reactors were built in a nuclear power station at Bradwell, Essex, which became operational in 1962 with an output of 300 MWe.

The pressure vessels are spherical in shape, 67 feet in diameter, in contrast to the cylindrical types at Calder Hall. Each graphite core, built from bricks into a 24-sided prism, is provided with 2,564 vertical fuel channels and 158 control rod channels. The fuel elements, 20,512 in each reactor, are natural uranium rods, clad in Magnox, 1.155 inches in diameter and 3.3 feet long.

Because the reactors were designed for the efficient generation of electricity they are operated at even higher temperatures than Calder Hall. The carbon dioxide coolant enters the reactors at 180°C and leaves at 390°C so that steam can be produced at 375°C. The coolant is also used at a higher pressure (127 psi) and faster flow rate than at Calder Hall, and the pressure vessels are provided with thicker walls to withstand the higher pressure.

A total of nine commercial Magnox power stations has been constructed in the United Kingdom, each

Trawsfynydd (above) the first inland British nuclear power station is in the Snowdonia National Park. The lake is warmed as it cools the 500 MWe Magnox reactor, causing ecological changes – resulting, in particular, in there being larger fish

One of the vertical cylindrical concrete vessels at Oldbury containing reactor, boilers, and blowers which is pre-stressed by layers of cables (right) carried in steel tubes laid in the wall in a helical pattern

Below: Locations of the nuclear power establishments in the United Kingdom

Dounreay 4

Hunterston 1, 5

Chapelcross 1
Hartlepool 5
Windscale 2
Calder Hall 1

Heysham 6

Wylfa 1

Connah's Quay 6

Trawsfynydd 1

Sizewell 1, 6

Berkeley 1
Bradwell 1

Oldbury 1, 7

Hinkley Point
Dungeness
1, 5   Winfrith 3   1, 5

1 – magnox reactor
2 – AGR
3 – SGHWR
4 – breeder reactor
5 – AGR (under construction)
6 – AGR (planned)
7 – HTR (possibly)

containing two reactors. The last two stations, completed at Oldbury, Gloucestershire, on the River Severn, in 1967, and Wylfa, in Anglesey, in 1970, have prestressed concrete pressure vessels which can take higher pressures than those made of steel.

Gas-cooled, graphite-moderated, natural-uranium reactors, very similar to the Magnox ones, have been built in France and bought from Britain by Italy and Japan. The total capacity of these reactors is about 8,000 MWe, of which over 5,000 MWe is generated by the British Magnox stations.

The extensive and rapid development of nuclear power by the United Kingdom before it had been demonstrated that nuclear stations were economically competitive with conventional ones was due to several factors. First, there was a large increase in the demand for electricity. Secondly, a considerable effort and a large financial investment had been made in nuclear energy for the nuclear weapons programme, culminating in the successful operation of Calder Hall, and many subsidiary industries and interests were involved. This had given reactor technology a certain 'momentum' and, consequently, it would have been difficult politically to interfere with its development to any appreciable extent. Thirdly, the United Kingdom Atomic Energy Authority, which had become a powerful body mainly because of the resources devoted to nuclear weapons, had eminent spokesmen anxious to argue the case for nuclear energy. Fourthly, this case was particularly strong in the mid-1950s because of a coal shortage which was expected to get worse in the future, and because of the perception that political instability in the Middle East might jeopardize the country's oil supplies. Finally, the United Kingdom has most of its population concentrated in a few highly industrialized areas and nuclear energy was regarded as a particularly suitable source of electricity. And it is probable that considerations of national prestige also played a not insignificant role.

But, in the event, experience was to show, within a decade, that nuclear stations are competitive with fossil-fuelled ones, at least in areas remote from large coal fields. And, because the United Kindom has a comparatively high fuel-cost economy, nuclear power – the source of 13 per cent of all the electricity now generated – has become an important factor in her economy. Thus, time has proved that the decisions leading to the rapid development of nuclear power by the United Kingdom were correct even though the original predictions, made during a wave of enthusiasm, were over-optimistic.

Reactor technology now advances so rapidly that the Magnox reactors are already obsolete. Future British nuclear stations will contain Advanced Gas

Cooled Reactors (AGRs) which will produce electricity more efficiently than the Magnox reactors – in fact, their efficiency will be about 42 per cent compared with about 33 per cent for the Magnox stations. AGRs use uranium oxide fuel elements consisting of clusters of rods canned in stainless steel, the uranium being slightly enriched in U-235. Fuel of this type has

A cutaway model of the 36-pin fuel element of an Advanced Gas-cooled Reactor. The construction of reactor fuel elements to the close tolerances necessary and in the very large numbers required is a major engineering enterprise

Dungeness 'B' showing the reactor core (red) in which pumps (white) force carbon dioxide over hot fuel elements, into the cavity (green) of the pressure vessel (yellow) and through the boilers (purple). Steam (blue) from the boilers drives the generator turbines and is recirculated

a longer life than earlier ones and, more importantly, allows the gas temperature to be about 200°C higher than in the Magnox reactors. At these temperatures, superheated steam can be produced at conditions suitable for the most modern turbine. As in the Magnox reactor, the moderator is graphite and the coolant is carbon dioxide.

Right: the aluminium-clad container vessel of the AGR at Windscale, and (opposite) the interior showing the top of the reactor core. The machine in the background is used for changing the fuel elements. Two heat exchangers can be seen to the right

These young kestrels were ringed for the ornithologists' record and replaced in their nest in a transmission tower at Dungeness which is built in a nature reserve. Nuclear reactors produce no pollutants and so have great siting flexibility

The core of the reactor and the heat exchangers are integrated in a prestressed concrete pressure vessel, whereas the Magnox reactor has external heat exchangers. This, and other safety factors, allows for great flexibility of siting so that AGRs can be installed close to large urban centres – the one being built at Hartlepools is within two miles of the city centre. And, because the design is relatively compact, the construction costs are much reduced.

An experimental AGR, constructed at Windscale, has been supplying 28 MWe to the national grid since 1962 and its very successful performance has encouraged the electricity authority to begin building three large AGR stations – the first at Dungeness, Kent, on the English Channel, alongside a Magnox station commissioned in 1965. The complete reactor,

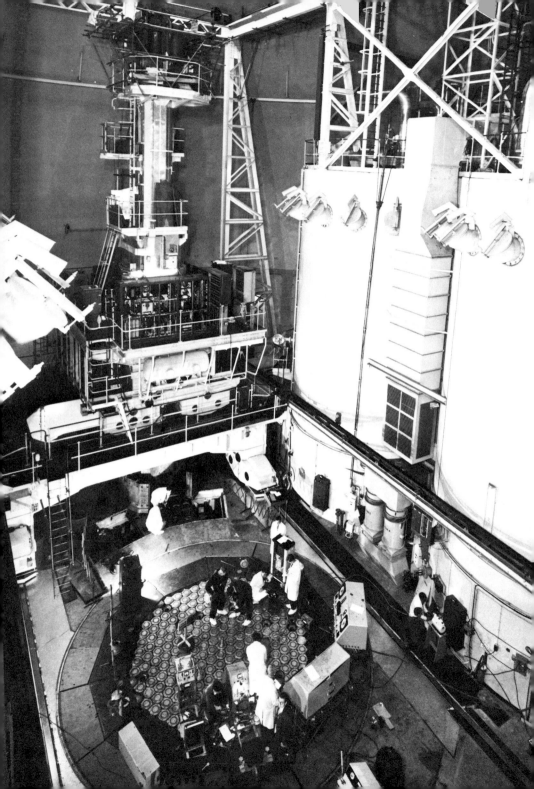

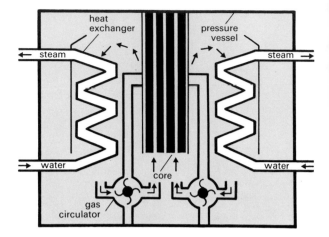

The circulation of gas in a gas-cooled reactor in which reactor, boilers and pumps are all enclosed by the pressure vessel

including the core, gas circulators and boilers, will be housed within a concrete pressure vessel, cylindrical in shape and internally clad with stainless-steel thermal insulation. The relatively small core, 27 feet high and 31 feet in diameter, contains 412 fuel channels each loaded with 8 fuel elements. The carbon dioxide is used at a much higher temperature and pressure than in the Magnox reactors. It enters the core at a temperature of 320°C and leaves it at 675°C, and the coolant gas pressure is 450 psi.

The most spectacular thing about the design of the Dungeness station is its compactness. It will contain two 600-MWe AGRs, the centres of which are only 160 feet apart, whereas the centres of the two 28c -MWe. Magnox reactors in the neighbouring station are 400 feet apart. The output of electricity per square foot of area of the AGR station is four times greater than that of the Magnox.

The Dungeness station should come into operation in 1973; and, by 1975, a total of 8,000 MWe will be being generated by five AGR stations. The fifth one, at Heysham, is planned to contain four AGRs generating a total of 2,500 MWe.

Light-water reactors
Light-water reactors have been developed mainly in

the United States; in them, ordinary water is used both as the moderator and the coolant.

At the present state of power-reactor development it is difficult to assess whether light-water reactors are more or less economic or reliable than graphite-moderated gas-cooled reactors. There is probably little to choose between them. Water cooling was originally chosen instead of gas cooling for the Hanford plutonium production reactors during the Second World War because the technology of this type was more advanced at the time, and its continued development into the light-water reactors was perhaps to be expected. But, had gas cooling been chosen, power reactor development in the United States might have taken a very different direction.

The main control-room at Dungeness which contains all the indicators and controls necessary for normal and emergency operation of both the reactors and generating plants. Any fault would immediately initiate the automatic shut down of the reactor

Pressurized water reactors require only two lifts to replace the central section containing fuel rods. Here U Thant examines a model

Light-water reactors are of two kinds: boiling water reactors (BWRs) and pressurized water reactors (PWRs). The BWR converts the cooling water into steam in the reactor. The water and steam are forced through the assembly of fuel elements by very large jet pumps and heated to a temperature of about 285°C at a pressure of 1,000 psi to drive a turbine. In the PWR, the pressure of the cooling water is kept at about 2,000 psi so that it can be heated to about 320°C without boiling. It is then piped to a heat exchanger in which water is converted into steam – at a temperature of about 245°C and a pressure of 540 psi – for the turbine.

The fascination of the BWR is that it is the most straightforward way of using the heat from a reactor. Because steam is fed directly from the reactor to the turbine, the loss of heat in an intermediate heat exchanger is avoided and the efficiency increased.

A famous BWR is the one used in the Dresden Nuclear Power Station located about 50 miles south west of Chicago, Illinois. It began operating in 1959 and has an output of 200 MWe. The physical size of the reactor is smaller than that of an equivalent Magnox. The pressure vessel is cylindrical, about 12 feet in diameter and 40 feet high. The walls, about 5.5 inches thick, are made of molybdenum-bearing carbon steel with an internal cladding of stainless steel. Four 22-inch nozzles are located at the bottom to pass in the cooling water and twelve 16-inch nozzles at the top to allow the steam to pass out.

The core of the reactor, cylindrical in shape, is 10 feet high and 11 feet in diameter. In it are 488 fuel elements, consisting of slightly enriched uranium dioxide pellets, half an inch in diameter, in Zircaloy tubing. Each fuel assembly contains 36 tubes in a 4.3-inch square Zircaloy box. Four such segments are joined to

Zirconium (above), much used in the construction of nuclear power reactors, is lighter than steel and highly resistant to corrosion and heat

Left: the 50-ton head of the Dresden boiling water reactor about to be bolted into place with 56 5-inch-diameter bolts

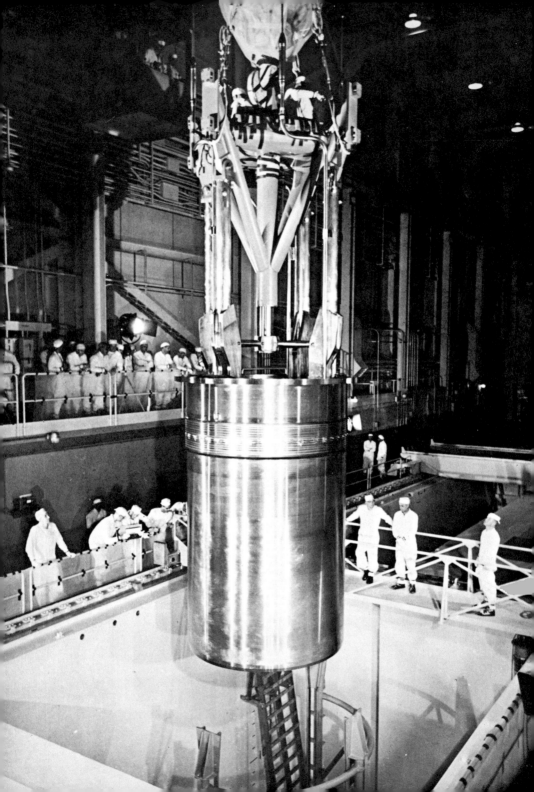

form an element 108 inches long. And the plant is confined in a steel sphere, 190 feet in diameter.

The American interest in PWRs has come about mainly from the experience gained in their development, during the 1950s, for ship propulsion. The first PWR to generate electricity, at Shippingport, Pennsylvania, on the Ohio River about 40 miles west of Pittsburgh, has been in operation since December 1957. In its first four years it produced over one billion kilowatt-hours of electricity. Water at a pressure of 2,000 psi is circulated through the core at 45,000 gallons per minute. It leaves the core through outlet nozzles at the top of the vessel, and passes to the steam generator. The pumps handle nearly 19,000 gallons per minute and are four times larger than any previously used.

The 160-MWe Yankee Nuclear Power Station, located at Rowe, Massachusetts, has been in operation since 1962. The core of the PWR contains 76 fuel elements held in place by upper and lower support plates. Each fuel assembly consists of 304 stainless steel tubes, 0.3 inches in diameter, filled with pellets of enriched uranium dioxide – each tube holding 150 pellets. The cylindrical core, 6 feet in diameter and 7.5 feet high, is much smaller than that required for a BWR of the same output. The pressure vessel is 9 feet in diameter and 31 feet high with walls 8 inches thick, clad on the inside with stainless steel. But the saving in cost from smaller size is offset by the need for separate steam generators and in fact, for a given output, the capital cost of PWRs and BWRs is about the same.

Countries other than the United States have installed light-water reactors. For example, the Soviet Voronezh Atomic Power Station, located on the Don River, about 25 miles from Voronezh, contains a PWR with an electrical output of 250 MWe. And several countries have purchased them from the United States. A total capacity of about 16,000 MWe – about 9,000 MWe of BWRs and about 7,000 MWe of PWRs – has been installed of which about 11,000 MWe are in the United States. The Americans have about 6,000 MWe of BWRs and 5,000 MWe of PWRs.

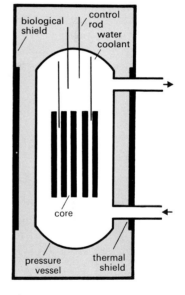

Above: a pressurized water reactor

Opposite: The 58-ton core of the Shippingport reactor poised over the reactor vessel. It contains 14 tons of natural uranium and 165 pounds of highly enriched uranium, enough to produce 100,000 kilowatts of electricity

Above: The Yankee PWR in Massachusetts has an output of 175 MWe and since 1960 has produced well over 10 million megawatt-hours of electricity. Opposite, top: the engine-room at the Voronezhsky power station. Opposite, bottom: India is developing the CANDU-type reactor for its own special needs. Nuclear power could transform this scene of reluctant muscle-power into one of all-year-round efficient farming with ample fertilizers and power and water for irrigation

An important feature of the American programme is the marked trend towards nuclear plants of larger size. Several plants of 1,000 MWe capacity are under construction or planned. The costs of constructing and operating any large power station are less than the total of those of two or more small plants giving the same total power, but this size advantage favours nuclear more than fossil-fuelled stations.

A second feature concerns the locations of the plants. At present, they are concentrated in New England, California, the North Atlantic Seaboard and the Great Lakes region, all areas of high fuel costs in which nuclear plants offer a clear economic advantage over conventional ones. Moreover, they are among the most highly populated and industrialized areas of the United States and, therefore, the demand for electricity is high. Other areas have high fuel costs, such as parts of the Midwest and the Pacific Northwest, but these

The distribution of nuclear power plants in the United States showing their concentration on the Atlantic sea-board – areas of high population density and high fossil fuel costs

do not have sufficient demand to justify nuclear plants of competitive size. And it is unlikely that nuclear power will be competitive in places like the Appalachian Mountains (coal) or Texas (oil and natural gas), or that it will be used to supply areas of low population density. In short, nuclear power is being extended to the major industrialized parts of the country (other than Texas) – areas which account for most of the electric power production.

The experience of the United Kingdom and the United States demonstrates that the optimum conditions for the nuclear generation of electricity occur in environments in which there is already a highly developed technology, a shortage of fossil fuels, a large electric power-grid network and large concentrations of population.

Heavy-water reactors

Heavy-water moderated power reactors are much less common than the other two types mainly because their development was inhibited by the scarcity and relatively high cost of heavy water. The substance is, however, an excellent moderator. Canada, in particular, has based her nuclear programme on these reactors.

The first heavy-water power reactor to start up was the Halden reactor at the Norwegian Institutt for Atomenergi, Norway, some 70 miles south of Oslo. This very small experimental reactor began operating in 1959. But the most famous one is the 208-MWe Canadian Deuterium Uranium Reactor (CANDU) located at Douglas Point, on Lake Huron in Ontario, Canada. The reactor, which uses heavy water as both moderator and coolant and natural uranium as fuel, began operating in 1966. The reactor core is contained in a stainless steel calandria tank, 20 feet in diameter and 17 feet long. The calandria is provided with 306 horizontal channels for the fuel elements. There are a total of 3,672 elements, each one a cluster of 19 tubes filled with natural uranium dioxide pellets. The coolant tubes run in the same channels as the fuel elements.

The reactor has a novel method of control, the level of the moderator in the calandria tank being varied. Four 500-MWe plants are under construction at Pickering Township, near Toronto and this complex is expected to expand to 4,000 MWe.

A heavy-water, gas-cooled reactor, of Soviet design, has been installed in Czechoslovakia, at Bohunice. The coolant is carbon dioxide at a pressure of 850 psi. This pressure is borne by a vessel, 62 feet high and 16 feet in diameter, having a wall thickness of 5.5 inches. The reactor will come into operation in 1971 with an output of 110 MWe.

Of the established types, the heavy-water moderated reactor is the most economical in its use of fuel. For example, the CANDU reactor requires, for a given power output, considerably less than half the initial fuel charge required by the others. And the annual re-

An aerial view of the nuclear complex at Pickering, Canada, which will be expanded to include several CANDU-type power reactors. With large uranium reserves, numerous nuclear scientists and uneconomically located fossil-fuel deposits, Canada's attractions to nuclear energy are great

plenishment is significantly less too. Moreover, it is possible to change from natural uranium fuel to one using thorium without shutting down the reactor or even reducing its power. Thorium is efficiently converted in the reactor into U-233, a fissile material which could be used to fuel other reactors. Since thorium fuel is likely to become as cheap as, if not cheaper than, uranium, the prospects for the HWR are good. India, which has enormous reserves of thorium, is very interested in these reactors and is installing two in a power station at Rana Pratap Sagar, with a total output of 400 MWe, and one at Kalpakkam near Madras, with an output of 200 MWe. Meanwhile, in Pakistan, a 125-MWe heavy-water reactor is under construction at Paradise Point near Karachi. The total capacity of heavy-water reactors so far installed is only about 2,200 MWe, of which more than one-half is in Canada.

Future developments

Because power reactor technology is continually surging forward, it is not surprising that the existing reactor types are being further developed in the light of accumulated operating experience. This is well illustrated by the gas-cooled reactors. Their size has increased from the 35-MWe reactor at Calder Hall to the 625-MWe reactors to be installed in AGR stations now under construction. Steam conditions have progressed, from the relatively primitive, to temperatures of 565°C and pressures of 2,300 psi – ideal for the most modern, high-efficiency turbogenerator sets. And the overall efficiency of the reactors has been increased from about 20 per cent to over 42 per cent. But the development will not stop there. The next step is to increase temperatures still further – to about 800°C. It may then be possible to operate gas turbines directly, thereby avoiding the heat losses which inevitably occur in heat exchangers. The production of these high temperatures is the aim of an international reactor research project, known as DRAGON, being undertaken at Winfrith, England, in cooperation with the OECD countries.

The helium-cooled, high-temperature DRAGON experimental reactor. Present studies indicate important economic advantages for HTR's especially with gas rather than steam turbines

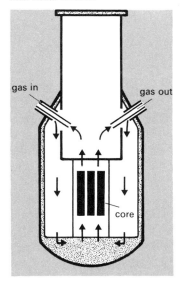

gas in

gas out

core

DRAGON is an experimental helium-cooled reactor with coolant temperatures ranging up to about 760°C. The generation of electricity is not planned and so the heat is discharged into the atmosphere. The fuel elements are encased in graphite, a new departure in fuel-element design. The graphite cladding also acts as the moderator. The fuel itself is a mixture of uranium, enriched in U-235, and thorium dicarbides. The thorium is converted to U-233 by neutron bombardment. Several different forms of fuel have been used. In one it is in the form of carbide rings, six of which are stacked in each of a number of graphite boxes, with a graphite rod passing through the centre of the rings. Seven boxes make up a fuel element and there are 37 elements contained in the core, which is 3.5 feet in diameter and 5.25 feet high. In another it is in the form of ceramic-coated pellets. The reactor has a somewhat complex shape, rather like a flask, with a diameter of nearly 12 feet, a height of 17 feet and a wall thickness of 2.25 inches. Concentric inlet and outlet ducts are provided for the helium coolant.

Uranium pellets to be used for producing fuel elements for the DRAGON reactor experiment

The reactor is contained inside two shells. The inner one, 60 feet in diameter and 80 feet high, is filled with nitrogen at atmospheric pressure. The outer shell, 110 feet across and 86 feet high, is made of concrete 2 feet thick.

At present, the higher temperatures produced by this type of reactor could not be used to full advantage. But there is little doubt that high-temperature helium turbines will soon be developed and these will have a much smaller bulk than the present steam turbines. The first commercial high-temperature reactor may be installed in a power station planned for Oldbury. This is to be of 2,500 MWe capacity, provided by four reactors, and will probably be commissioned soon after 1975, possibly with newly-developed gas turbines.

Improvements in the performance of light-water reactors will be achieved by advances in fuel-element design leading to lower fabrication costs, longer life and improved performance. More suitable cladding

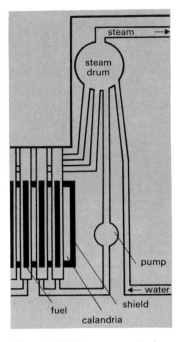

The SGHWR represented above uses heavy water as moderator and light water as coolant

Opposite: incandescence from the heat exchanger tubes of a molten-salt reactor in which the salt is cooling from 650°C. Air is passed over the tubes to dissipate the heat

materials, such as iron-aluminium alloys, are being developed and tested. And the costs of reactor construction will be reduced by the use of materials such as carbon steel instead of stainless steel.

An interesting new type of heavy-water reactor is the Steam Generating Heavy Water Moderated Reactor (SGHWR), a prototype of which has been constructed at Winfrith, England. The reactor began operating in 1968 and generates about 100 MWe. The core consists of a bank of 112 vertical zircaloy pressure tubes in a calandria containing the heavy-water moderator. The cylindrical calandria, made of aluminium alloy, is 12 feet in diameter and 13 feet high. Each pressure tube contains one 12-foot-long fuel element. The heat produced in the fuel is removed by light-water coolant, which passes upwards through the tubes and which is partially turned into steam. The mixture of water and steam is then passed to two drums where the water is separated from the steam. The water is pumped back to the core and the steam fed directly to the turbine. The light water acts as a second moderator as well as the coolant.

The fuel is slightly enriched uranium in the form of pellets, 0.57 inches in diameter, contained in tubes of Zircaloy. A cluster of 36 tubes makes up each fuel element. The reactor is controlled by varying the level of the moderator in the calandria.

A commercial SGHWR would be essentially similar to the Winfrith prototype. Using this unique design, reactors of different outputs can be constructed from standard size channel-tube assemblies and fuel elements. The reactor is, therefore, likely to be economic in small sizes and of considerable interest to those developing countries which are about to introduce nuclear power.

A number of other reactor types have been toyed with, but none has yet proved as successful as the established types. The most interesting example is the molten salt reactor. The advantage of this type of reactor is that, because the fuel is in solution, the fabrication of fuel elements is not necessary. And, since the

An aerial view of the steam generating heavy water reactor at Winfrith. This type of reactor is of great interest to developing countries because it can be made to be economic in small sizes

reactor operates at high temperatures, it should have a high efficiency. Molten salt fuels were originally developed for use in compact, high-temperature, high-power reactors suitable for aircraft propulsion. Only two reactors of this type have been built and operated. One was the Aircraft Reactor Experiment (ARE) at Oak Ridge National Laboratory and the second, now operating at Oak Ridge, was built to investigate its potential for commercial power plants.

The core of the ARE, which operated briefly in 1954 and was then dismantled, was made of blocks of beryllium oxide stacked to form a cylinder, only about 3 feet in diameter and 3 feet high. It was because of its small size that it was hoped that the reactor could be used to power aircraft. The fuel was a mixture of sodium, potassium and beryllium fluoride containing uranium hexafluoride. The total mass of U-235 in the system was 66 kilograms. The molten salt fuel was

circulated through nickel alloy tubes arranged verti-
cally in the beryllium oxide moderator. The fuel,
which also acted as the coolant, circulated at a high
temperature, entering the core at 650°C and leaving it
at 820°C. The nickel alloy pressure vessel was also
small, 4 feet 4 inches in diameter and 3 feet high. The
reactor ran successfully for about 10 days before it was
dismantled.

The Molten Salt Reactor Experiment (MSRE) is
housed in the building used for ARE. It has a cylindri-
cal graphite core, about $4\frac{1}{2}$ feet in diameter and $5\frac{1}{2}$ feet
high. The molten salt fuel is a mixture of lithium,
beryllium, zirconium, fluorine, thorium and uranium.
There are about 600 channels in the core through
which the fuel flows at a rate of 1,200 gallons per
minute. There are staunch advocates of the molten
salt reactors, but serious problems remain to be solved
and their development will be slow.

*At a lonely site on a desert plain the
Aircraft Research Reactor towers
over two railway flat-cars pushed on
parallel sets of tracks by a shielded
locomotive (left). The reactor (centre)
heated the air fed to it by the intakes
of the turbo-jet engines (right) which
were themselves powered by this
heated air*

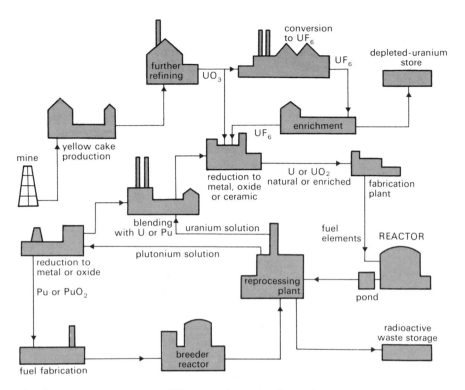

The following labels appear in the diagram:

conversion to UF$_6$

depleted-uranium store

further refining

UO$_3$

UF$_6$

enrichment

yellow cake production

mine

UF$_6$

reduction to metal, oxide or ceramic

U or UO$_2$ natural or enriched

fabrication plant

blending with U or Pu

uranium solution

fuel elements

REACTOR

reduction to metal or oxide

Pu or PuO$_2$

plutonium solution

reprocessing plant

pond

radioactive waste storage

fuel fabrication

breeder reactor

The cycle of processes involved in the manufacture of nuclear fuel. Because it is rare for all these industrial plants to be located in one country transport of radioactive material by land and sea is frequently necessary

The nuclear fuel cycle

In countries with advanced nuclear power programmes – the United States, the United Kingdom, the Soviet Union, France, and the German Federal Republic – large nuclear industries have been established. For example, in the United States the electric utilities may have over 90 plants installed by 1975, representing an investment of over $10 billion, a figure which may well be doubled by 1980. And, if the predicted nuclear capacity is, in fact, installed, the annual fuel costs will alone amount to about $2 billion by 1980. Nuclear industries will grow to giant sizes during the 1970s and will be characterized by the advanced technology and ruthlessness which exist in, for example, the oil and aviation industries.

The nuclear industry can be divided into two sections – one designing and installing power reactors and the other providing the fuel for them. The latter is based on the nuclear fuel cycle. A cycle occurs because nuclear energy creates quantities of new fuel at the same time as it consumes fuel. The new fuel can, in turn, be used as a source of nuclear energy, creating still more fuel, and so on. In today's reactors, the original fuel is contained in uranium. As uranium 'burns' in the reactor, plutonium is produced, and this will be the fuel for future reactors. As plutonium burns it can produce more plutonium in uranium, and so on. But the starting point of the cycle is virgin uranium.

After mining, the uranium ore is refined and, if necessary, enriched in U-235. It is then fabricated into fuel elements and sent to the reactor. After use, they are taken to a reprocessing plant where the uranium, plutonium and the fission products are separated. The disposal of the highly radioactive waste from this plant presents special problems. Many of these processes are complex, requiring large industrial establishments.

There is no major difficulty in mining uranium ores. Conventional underground and open-pit methods, and place-leaching are used. After mining, the ore is processed in mills to extract the uranium in the oxide form. Because of its colour, the miners call this material 'yellow cake'. After production, yellow cake is refined and purified, and dealt with in one of two ways, depending on whether it is wanted to provide fuel or a

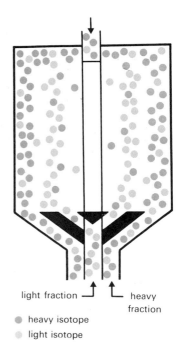

light fraction — heavy fraction

● heavy isotope
● light isotope

Above: A gas centrifuge. A mixture of two isotopes of nearly equal mass is admitted at the top. The centrifuge spins at 50,000 rpm throwing the heavy molecules to the edge so that the lighter may be drawn from the centre. Left: A centrifuge cascade. The light and heavy fractions are cycled in opposite directions and accumulate in separate reservoirs

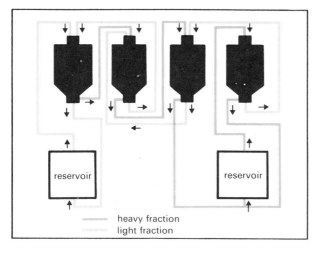

reservoir   reservoir

—— heavy fraction
—— light fraction

reactor burning natural uranium (a Magnox or heavy-water reactor) or one burning enriched uranium (an AGR or light-water reactor).

The gas-diffusion method is, at present, the only one actually used to produce uranium enriched in U-235 on a large scale for reactor fuel. Eight diffusion plants now exist: three in the United States, two in the Soviet Union, and one in each of Great Britain, France and China. These plants were all originally built for military purposes, but they are now mainly used to produce enriched uranium for power reactors. No plant has been built outside the nuclear-weapon countries. The degree of enrichment required is low – normally less than 4 per cent. The details of the gas-diffusion method have already been described, in chapter 1.

Several countries are attempting to develop an alternative method of uranium enrichment. In particular, development work on the gas centrifuge has begun in the United States, Great Britain, South Africa, West Germany, the Netherlands, Japan and Australia. The centrifugal method of separating isotopes in gaseous form is based on the principle that the gravitational force on a particle is proportional to its mass. The suggestion that this could be applied to the separation of isotopic molecules of different masses was made by F. A. Lindeman and F. W. Aston as long ago as 1919. They suggested that significant separation could be achieved by the use of a centrifuge, which provides a field of force analogous to gravity but much more powerful.

In essence, a suitable gas centrifuge for uranium enrichment would probably consist of a vacuum tank containing a long, rotating pipe-like drum with concentric nozzles at one end and concentric orifices at the other. Uranium hexafluoride gas would be pumped in via the nozzles and, as the gas moves up inside the rotating drum, molecules would tend to be flung outwards. But, because a pressure gradient builds up, molecules of the lighter U-235 isotope would tend to diffuse towards the centre. In practice, the gas would move through the centrifuge in two streams and, under equilibrium, there would be no net mass transfer between them. This means that the number of molecules of each mass flowing in each direction is slightly different; the inner stream thus becomes enriched in U-235 and this is collected by the inner exit orifice. This slightly enriched flow would be fed to the inner nozzle of the next centrifuge in the cascade and so on. Although a separative factor per stage about 10 times higher than that for the diffusion process can probably be achieved, a centrifuge plant would still require a very large number of centrifuges in cascade to provide a useful output of enriched uranium.

Throughout its history, the gas centrifuge has been dogged by materials problems. The limiting factors governing performance have been the tensile strength and density of the material of the outer casing of the drum and the rotor bearings. Drum speeds of about 450 metres per second are now believed possible and, therefore, a centrifuge rotating at 60,000 r.p.m. will have a diameter of about 14 cm. Such a centrifuge is likely to be about one metre long. These parameters rule out the nickel and aluminium alloys used to handle uranium hexafluoride in diffusion plants. Conventional steels seem little better. And it is likely that the current developments in gas centrifuge technology, mainly in Britain, Holland and West Germany, are largely the result of a somewhat revolutionary material having become available. The material referred to is possibly carbon fibre. In this case, strength could be supplied to the outer wall of the rotor using a filament winding technique, with resin-coated carbon fibre being wound continuously round a mandrel of rotor diameter. This fibre case could then be pushed over a liner of aluminium resistant to uranium hexafluoride.

Where the centrifuge scores over the diffusion process is in running costs. The amount of electricity required to operate a centrifuge installation is said to be one-fifth (some say one-tenth) that needed to power an equivalent diffusion plant. The capital costs are about the same, but the structures of the two investments are different. While a diffusion plant needs a massive initial injection of capital for the thousands of stages before enriched uranium can flow, a working centrifuge plant can be economically built up step by step as demand for enriched uranium increases.

The enriched uranium hexafluoride is converted either into uranium dioxide (UO_2) powder or into the metal form to produce reactor fuel. If natural uranium fuel is required, the purified yellow cake is directly converted into UO_2 or metal. Most modern reactors employ fuel in the form of uranium dioxide; this is often sintered into pellets and then sealed in tubes.

The fabrication of reactor fuel elements requires a highly developed materials technology. The elements must be able to withstand extremely arduous operational conditions. They must be encased in materials which do not react with the reactor coolant, and which will withstand the large quantities of heat generated in the fuel elements and the effects of high levels of radiation. Close dimensional tolerances are necessary to maintain proper heat transfer to the coolant. Moreover, some of the fission products produced within the

Above: uranium oxide pellets. Below: a laser beam punctures a can so as to test the fission products of an irradiated fuel sample

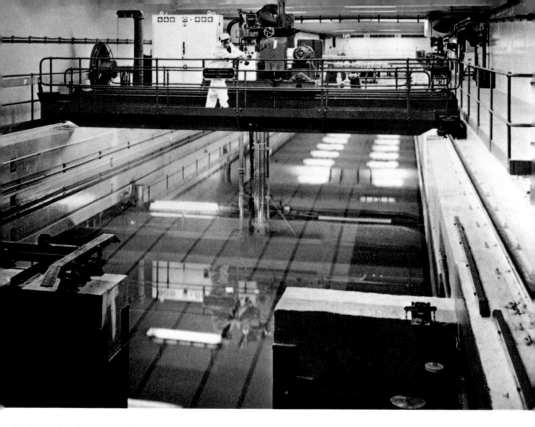

fuel can lead to complications. For example, about 30 per cent of the fission products are gases such as krypton and xenon, and these collect as bubbles in the fuel, causing it to swell and eventually to rupture the can.

A given fuel element may yield heat continuously for up to 3 or 4 years before it is removed from the reactor. Even after this time, only about 1 per cent of the fuel is burned up. But the element must be removed because many of the fission products will absorb neutrons. The value of the multiplication factor will, therefore, steadily decrease as more fission products are produced and eventually fall below 1, whereupon the chain reaction will cease. Moreover, as plutonium accumulates it competes with U-235 for thermal neutrons and eventually as much plutonium would be lost by fission as is produced. Since plutonium is required for fuelling future reactors, the time at which the fuel element is removed from the reactor is a compromise

The Bradwell cooling pond. Several months in the pond considerably reduces the radioactivity of fission products in irradiated fuel elements which are then reprocessed at Windscale

A flask of irradiated fuel elements being moved into a shielded cell at Windscale where the elements are taken from the flask for storage in a cooling pond

between obtaining the maximum amount of plutonium and the necessity of maintaining the chain reaction.

When the spent fuel elements are removed from the reactor they are transferred to a 'cooling pond' of water for a few weeks to allow the U-239 to convert to Pu-239 and to permit most of the highly radioactive fission products to decay. The elements are then transported, in heavily shielded containers, to a reprocessing plant where the unused uranium and the plutonium, which are valuable assets, are removed.

A reprocessing plant is a complex and costly chemical establishment and, because the capital cost is relatively independent of the capacity of the plant, economic reprocessing can only be achieved if a large-scale plant is used to serve many reactors. In the plant, the fuel elements are first decanned and are then dissolved in acid. The acid solution is fed into a multicycle solvent extraction system in which the uranium and plutonium are separated from each other and from the fission products. The uranium and plutonium are then purified to a degree depending upon the ultimate use of the materials. The recovered uranium, for example, can be blended with fresh uranium, to restore the level of enrichment, and refabricated into fuel elements. The efficiency of the recovery of the fissile materials is normally nearly 99 per cent.

Disposal of radioactive waste

Nuclear reactors produce, by pre-1939 standards, enormous amounts of radioactivity. For each megawatt of electricity generated by a reactor, about 3 grams of uranium are consumed daily and the same amount of fission products is produced. Initially, the radioactivity associated with a gram of fission products is very high indeed, but it decays away rapidly to begin with and then more slowly. The quantity of radionuclides present in a radioactive sample is measured in a unit called the curie. One curie corresponds to the amount of a radionuclide in which the number of disintegrations is 37,000 million per second. Originally, the curie was defined as the amount of radioactivity associated with one gram of radium – hence the rather odd number. Even after 100 days, about the time which usually elapses before the fission products are separated from the fuel elements, the radioactivity of a gram of fission products is still about 500 curies. The British power reactors alone are producing about 20,000 grams of fission products per day which, even after 100 days decay, amount to several millions of curies per day. And in one month's output of fission products there is a sufficient amount of the hazardous radionuclide strontium-90 to provide the maximum body burden allowed by the International Commission for Radiological Protection, the organization which sets the safety standards for the use of nuclear energy, for the entire world's population.

The dispersal of these huge amounts of highly-radioactive materials into the oceans or the air is quite

Right: tanks at Windscale used for storage of liquid wastes until their radioactivity has decreased to an acceptable level for disposal at sea. Above: the interior of a storage tank buried deep underground for long term storage of high level radioactive wastes. Heat produced by radioactive decay is removed by the cooling coils shown here

unacceptable and impracticable, and so they have to be stored. Most of the fission products are contained in the waste solution from the extraction columns of the reprocessing plant. This is first concentrated in as small a volume as is possible and then stored in a manner which will ensure that a leakage of radioactivity cannot occur under any circumstances and with sufficient shielding to reduce the radiation in the storage areas to safe levels.

Exotic methods have been suggested for disposing of radioactive wastes, such as storage in uninhabited areas – like the polar regions or deserts – or firing them into space but these are far too expensive to be used and, in practice, more mundane methods are adopted. In Britain, the high-level radioactive waste from the reprocessing plant is stored indefinitely in underground storage tanks at Windscale after the bulk of the solution has been reduced as far as possible by evaporating off the excess water. The tanks, fabricated from half-inch thick stainless steel and surrounded by thick concrete, are virtually indestructible and are sited far enough underground to be adequately shielded.

But wastes of much lower radioactivity arise from many of the processes in the plant and very considerable volumes of these are produced. It would be uneconomical to evaporate these down for storage and they are, therefore, discharged at sea through effluent pipes which run out well past the low-water line. The amount of radioactivity disposed of daily in this way is strictly controlled and frequent tests are made to ensure that there is no dangerous build-up of radioactivity. This is necessary because the radioactivity could become reconcentrated by, for example, deposition on the sand or mud of the sea-bed, possibly leading to contamination of fishing gear, or by uptake in algae on which fish feed.

Even though the storage of radioactive wastes as liquids will provide a suitable means of dealing with the problem for several years eventually other methods will have to be found. The best would be the reduction of the wastes to a solid form, possible by its incorporation into a suitable clay-like material which could then be hardened.

Scientists lower a camera into an empty waste-tank to inspect the interior. Left: the radioactivity in liquids can be trapped in clay and fixed by baking in preparation for permanent disposal in abandoned salt mines or underground vaults in stable crystalline rock

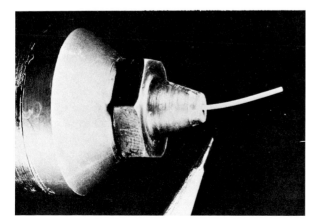

The rapid acceleration in the use of nuclear power has exceeded all expectations, even though it is hardly surprising in view of the extraordinary increase in the demand for electricity. In most industrialized countries, demand is doubling about every ten years and will continue to do so for at least another 20 years. And, in many of the developing countries, the doubling time is considerably shorter.

The experience of the United States dramatically illustrates what will sooner or later become the general pattern. The use of electricity is growing at a rate twice as fast as the Gross National Product and five times as fast as the population. And the annual orders for electricity generating capacity tripled between 1964 and 1968. Because of the rapid rate of technological advance, new uses for electrical energy are being continuously developed. The rapidly growing use of air conditioning, which in the past few years has produced large additional demands for electricity, is a typical example. And, as the society becomes increasingly automated, so the demand for power soars. Furthermore, several power failures during the recent past have driven home the smallness of the margin between generating capacity and peak demand – as demand increases so will the need for assured and adequate power reserves. The electric utilities must, therefore, constantly add to their generating capacity. Over the past two years, the amount of new generating plant

The two largest pressure vessels ever built are, for maximum efficiency, linked by the common control room of the 1,180 MWe Magnox power station at Wylfa in North Wales

ordered and scheduled was equal to over one-half of the total installed capacity.

The increase in the demand for electricity is, of course, a direct reflection of the explosive increase in demand for energy in general. But the increase in per capita energy consumption up to the year 2000 will vary from region to region. Because energy comes from several sources, it is normal practice to measure its consumption in tons-of-coal equivalent. In North America, the per capita consumption in 2000 is likely to be about 15 tons-of-coal equivalent, compared with about 8 tons in 1960. In Western Europe, the corresponding figures are about 5.5 and 2.7; in the USSR and Eastern Europe, 7.5 and 3; in Japan, 4 and 1; in Latin America, 1.7 and 0.7; and in Africa, 0.4 and 0.3. For the world as a whole, the per capita consumption in the year 2000 will probably be about 3 tons-of-coal equivalent, whereas in 1960 it was 1.5 tons.

The population of the world in 2000 may be 7,000 million and the overall consumption of energy nearly 25,000 million tons-of-coal equivalent. In many areas, supplies of fossil fuels will be inadequate to meet these enormous demands and it is most likely for this reason that the authorities are looking to nuclear power. But it must be emphasized that the success of nuclear power does not depend on the threat of a future shortage of fossil fuels or an increase in their prices. Coal and oil have had relatively stable prices for a long time and, in general, there are large reserves of them. Power reactors are being installed because it is believed that economic and other benefits can be gained from their use.

But the extent of the future development of nuclear power will depend upon the adequacy of nuclear fuel supplies and the trends in the costs of nuclear electricity. These two factors, to be discussed in this chapter, are inter-related since a shortage of nuclear fuels will inevitably lead to an increase in nuclear costs.

Benefits and costs: United Kingdom

From the experience so far gained from the more advanced nuclear power programmes it is possible to assess realistically the benefits and the social costs that accrue from them. The main justification for nuclear power is that nuclear electricity is cheaper than electricity generated by fossil-fuelled plants. It is therefore important to consider the circumstances in which nuclear plants are competitive and exactly how competitive they really are. Unfortunately, no simple answers can be given to these questions.

The factors determining the cost of nuclear power are: the capital cost of constructing the nuclear power station; the fuel costs; and the power-station operating costs. From these figures the generation costs of electricity can be determined. The major component is the charge on the capital cost of the nuclear station, which amounts to 60 or 70 per cent of the total. The fuel costs account for about 30 per cent and the operating costs less than 10 per cent.

The capital costs of the commercial nuclear stations built in the United Kingdom are decreasing as more stations are constructed and as the size of reactors increases. Thus, the cost of the first nuclear station at Berkeley (output 320 MWe), commissioned in 1962, was £185,000 per MWe whereas the cost of the latest nuclear station at Wylfa (output 1180 MWe), commissioned in 1969, was £110,000 per MWe. Allowing for the fall in the purchasing power of the pound in the interval, the reduction in real terms was about 50 per cent. But, on average, the capital cost of a Magnox station was, for a given output, about twice that of a conventional station.

The costs of AGR stations, however, will be significantly less than those of Magnox stations. For example, it is estimated that the total capital cost of the Hartlepools AGR station will be about £100,000 per MWe. The cost of a coal station of the same capacity built at Hartlepools has been estimated at £71,000 per MWe.

These figures demonstrate that the gap between the capital costs of nuclear stations and those of conventional ones is narrowing significantly with time as the costs of the former decrease and the costs of the latter

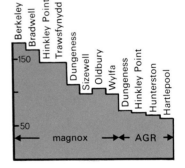

Since the construction of Berkeley, the first British commercial nuclear power station, capital costs have steadily been reduced and are now approaching those of fossil-fuelled stations

increase. The running costs of a nuclear station are only about one-third of those of a coal-fired station, the exact fraction depending on the type of reactor and the relative prices of coal and uranium. The annual fuel replenishment of a reactor is only about a quarter of a ton of uranium per MWe whereas a coal-fired station requires annually about 3,000 tons of fuel per MWe.

To calculate the generation costs of electricity it is necessary to establish the 'ground rules' on which the estimates are based. Firstly, a value must be decided for the load factor, or availability, of the power station, defined as the percentage of the maximum possible yearly usage. For nuclear plants it is usual to take an average load factor of 75 per cent which means the full generation of electricity for $273\frac{3}{4}$ days in the year. Secondly, a period over which the plant will operate must be assumed – a value of 20 and 25 years is normally taken for the lifetime of a Magnox and AGR respectively. Thirdly, a value for the burn-up factor of nuclear fuel is required; this is defined as the total amount of heat released per unit mass of fuel and is usually expressed in megawatt-days per tonne (MWD per Te). Finally, capital charges must be calculated on the basis of the annuity appropriate to an assumed rate of interest (normally 8 per cent) and the lifetime of the reactor.

In normal practice, the unit of generation costs used is the 'mill' per kWh sent out by the station. A mill is one-thousandth part of a United States dollar or one-tenth of a penny at present exchange rates. On realistic ground rules, the cost of generating electricity from Magnox stations has fallen from 12.5 mills per kWh for the Berkeley Station, to 7.0 mills per kWh for Wylfa.

The generation costs for the first AGR stations will be 5.6 mills per kWh for Dungeness B and 5.2 mills for Hartlepools. For comparison, the generation costs from recently commissioned coal-fired stations vary from 5.4 mills per kWh for Ratcliffe to 7.4 mills for Aberthaw B and the generation cost from a recently

British nuclear electricity generation costs in mills per kilowatt-hour showing that nuclear fuel is competitive, particularly in areas remote from sources of fossil fuel

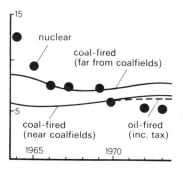

commissioned oil-fired station (Pembroke) is 5.8 mills – if oil were not taxed this would fall to 4.7 mills.

The capital cost of AGR stations can be expected to decrease, as the number built increases, due to the introduction of larger reactors and the spreading of overhead costs. And further reductions will result from technical improvements and the replication of design. In the period 1970–5 capital costs of AGR stations will probably decrease from about £74,000 per MWe to about £63,000, excluding interest during construction and the initial fuel cost. And generation costs will probably decrease during this period from 5.2 to about 4.6 mills per kWh.

Nuclear generation costs are likely to decrease further after 1975 – between the period 1976–80 they will probably fall well below 4 mills per kWh. But the actual costs will depend on the size of the nuclear programme. Advantages from increase in size, replication and technical development could, if fully realized, reduce capital costs to almost one-half of the present values. And there is likely to be a similar decrease in the initial fuel costs. Conventional generation costs during this period may fall to about 4.2 mills per kWh for untaxed oil, 5.1 mills per kWh for taxed oil and 5.8 mills per kWh for coal, assuming that the costs of these fuels remain the same as at present.

United States

The estimated generation costs of electricity from American nuclear stations have fluctuated considerably. In 1964, the General Electric Company forecast for the Oyster Creek plant a capital cost of $134,000 per MWe for a total capacity of 515 MWe. It was claimed that the capacity of the plant would, in practice, be about 640 MWe, thus reducing the cost to $108,000 per MWe. And the generation costs were calculated to total 3.6 mills per KWh, made up from capital charges of 1.5 mills, fuel costs of 1.6 mills and operating costs of 0.5 mills. The prices were calculated to give a marginal profit if no repeat orders were obtained, but a reasonable profit if a number of similar

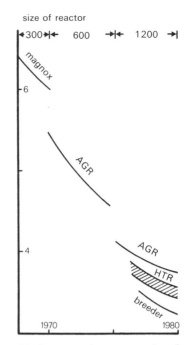

Nuclear generation cost – trends and predictions. Nuclear fuel costs promise to decrease more rapidly than fossil-fuel costs

plants were ordered. And, in 1966, the Tennessee Valley Authority calculated the cost of power from two 1,064-MWe BWRs would be 2.37 mills per kWh compared with 2.83 mills for a coal-fired station. These figures were obtained using more favourable ground rules than those taken in the United Kingdom. In contrast, the quoted cost of a 1,000-MWe PWR station at Diablo Canyon was about $180,000 per MWe.

These figures illustrate the large differences in the capital costs of reactors built at different sites in the United States by different utilities. In the south-eastern part of the United States, for example, costs are considerably less than elsewhere, mainly because labour costs are lower and supplies of cooling water are easily accessible.

An appendix published in 1967 to the 1962 AEC Report to President estimated that the cost of electricity from a typical 500-MWe nuclear plant had fallen from about 6.2 mills per kWh to 4.4 mills between the years 1962 and 1967. And it was forecast that there would be a further decrease, to between 3.5 and 4.2 mills by 1970. But this proved to be optimistic and a more realistic figure for present generation costs, based on similar ground rules to those used in the United Kingdom, is 4.5 mills per kWh.

The estimates of nuclear generation costs have been challenged by some authorities. And the fact that the initial impetus of nuclear technology was provided by military programmes has been quoted as a complicating factor. For example, all the uranium enrichment plants, the early reprocessing plants and the plutonium-production reactors were originally built for military purposes. It has been claimed that this effectively subsidized the peaceful nuclear energy programmes in the nuclear-weapon states. And governments are said to have actively encouraged, and overstated the case for, nuclear energy in order to maintain and strengthen the technological base necessary for the acquisition or the continued development of nuclear weapons. Because of the antagonism of the public towards these weapons it has been argued that some peaceful nuclear pro-

Opposite: the American nuclear power reactor at Oyster Creek, New Jersey, during construction.

grammes were used, irrespective of any economic justification, as a shield behind which military programmes proceeded. In this context, the price at which enriched uranium has been, and is, sold, has been questioned since governments may keep this artificially low. There may be some truth in these arguments in so far as the earlier nuclear power programmes were concerned. But it is also true that the recent adoption of nuclear power by many of the non-nuclear-weapon states indicates that it has now become an attractive economic proposition.

The fuels with which nuclear power competes are supplied by large and powerful industries whose labour and management have attempted to prove that nuclear power is not economically competitive with power generated by their own particular product. This has resulted in heated debates concerning the last few per cent of the costs of electricity generation. From the individual consumer's point of view it seems rather pointless to argue over differences of as little as a few hundredths of a penny in unit generation costs when the distribution and other costs of electricity bring up the price he pays to 2 or 3 pence per unit. But considerable sums may be involved for the industry as a whole. Thus, in the United States, where about 2.7 trillion kWh of electricity will be generated in 1980 a saving of 1 mill per kWh on generation costs would equal $2.7 billion per year.

There are important benefits to be gained from nuclear power in addition to economic ones and, even if it were marginally more expensive, these other benefits would make it advantageous to install it in many locations. But it is understandable that arguments have arisen over the pace of nuclear power programmes since there are definite social costs to be paid. There is, however, no question that nuclear fuels will replace coal, oil or natural gas in the foreseeable future: the demand for energy is increasing so rapidly virtually everywhere that each type of fuel has an important role to play. And it is in the best interests of the national economy that all forms of energy should be allowed to operate in conditions best suited to their continuing development. Once this is realized, many of the often emotive arguments used against nuclear power lose their force.

Other considerations

One important advantage of nuclear plants is their great siting flexibility. For example, large nuclear stations are being constructed within ten miles of a city centre in Canada, and in the suburbs of Leningrad. Fossil-fuelled stations are normally sited near coal fields or oil refineries. Nuclear power can have beneficial effects in remote or depressed regions by stimulating the introduction of new energy-intensive industries.

A nuclear programme encourages the development and application of many scientific and technical skills, and this spills over into the commercial use of products and processes. Many techniques, e.g. welding and materials purification, greatly benefit from a nuclear programme. And the physical sciences and technology are encouraged to develop to the most advanced levels.

Obtaining fuel for nuclear plants requires far less labour than for coal-fired plants – an important consideration in areas of labour shortage. For example, the fuel for an AGR of 1,000 MWe capacity (about 200 tons of uranium per year) is obtained by about $\frac{1}{8}$th of the manpower required to obtain the coal needed (at least $2\frac{1}{2}$ million tons per year) for a coal-fired station of equivalent capacity. The number of persons employed at large modern power stations is roughly the same for all types – less than one man per MWe capacity. But nuclear power stations require a larger labour force for construction.

Although nuclear stations at present need more capital for construction there is a high internal rate of return on the extra capital. And nuclear capital costs are being reduced, whereas conventional costs are increasing, so that the gap between them is narrowing.

Accident rates in nuclear stations are lower than in conventional stations. There are no high-level bunkers, fewer cat-walks and staircases in a nuclear power station and so the accident rate from falls is relatively low. In the nuclear energy industry all reasonable precautions are taken to protect workers against radiation hazards. The extent of these hazards was fully appreciated when the industry was first set-up, and codes of practise prepared before nuclear stations were constructed. This has rarely happened in other industries.

The nuclear industry has a good public health record. The effects of oil-tanker accidents can be widespread and the bulk transport of fuel by road involves some risks. But, because uranium fuel is far less bulky than other fuels, the transportation problems associated with it are less serious.

Although the discharge of waste from conventional power plants is being steadily reduced, the burning of fossil fuels releases relatively large quantities of pollutants into the atmosphere and, in some areas, this has already caused a serious health hazard. The discharge of radioactive gases from nuclear stations is usually

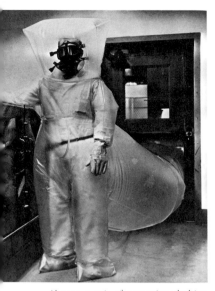

Above: a suit of protective clothing with clean air supplied through the access tunnel

strictly controlled and is of no environmental significance. It should be noted that coal, oil and natural gas
contain small quantities of naturally-occurring radionuclides and the combustion of these fuels involves
the release of quantities of radioactivity into the atmosphere. And, in fact, fossil-fuelled power stations may
be discharging relatively larger amounts of radioactivity than nuclear stations although the amounts
involved are, of course, also insignificant. The effluent
of high radioactive content from nuclear stations is
relatively small in quantity, and can be stored in tanks.
All of the highly radioactive liquid waste from the
British reactors up to the mid-1980s could be accommodated in a number of tanks occupying less than an
acre of land. In some cases, however, radioactive
effluent is directly discharged into rivers, for example,
and there is then a danger of a deleterious ecological
effect on the local environment. And as the use of
nuclear power becomes increasingly widespread the

*The transportation of nuclear fuel is a
simple operation compared to that of
fossil fuels. Above: fuel from a
Canadian reactor en route to the
United Kingdom for reprocessing*

*Opposite: irradiated fuel from Garig-
liano is neatly borne by rail to
Windscale. The fuel in these containers is equivalent in energy production to several trainloads of fossil
fuel*

The clouds rising from Calder Hall nuclear power station contain no radioactive material or other contaminants; this is water vapour from the cooling towers

control of the disposal of radioactive waste will become a serious problem, the solution of which will undoubtedly require international collaboration.

Nuclear power reduces reliance on oil although, in some countries, supplies of other fossil fuels are under national control. Where fossil fuels are indigenous and there are no local supplies of cheap uranium the use of nuclear power will, of course, involve the expenditure of foreign exchange.

Nuclear power lessens the effects of industrial disputes on electricity supplies because reactor fuel needs replacing relatively infrequently and is easily transportable.

Organic substances are needed in large and increasing quantities for the rapidly expanding chemical industry and it is important that the use of accessible supplies for fuel should be minimized. Nuclear power is clearly beneficial in reducing the rate at which organic substances are used up.

There are problems common to all forms of power production. The local heating of the environment of a power station could produce adverse ecological changes – a nuclear station discharges more waste heat than does a conventional one. Strip-mining to obtain cheap coal and, possibly, low-grade uranium ore, spoil large areas of landscape. And, as with all forms of modern technology, power production produces undesirable effects on the physical environment.

Nuclear fuel supplies

There is some uncertainty about the future fuel supplies for reactors. Uranium is a widely distributed element found in a large variety of minerals. Economically mineable deposits occur in sandstones, shales, granites, phosphates, lignites, quartz-pebble conglomerates and veins. But most of the uranium is dispersed through the rocks of the earth's crust and only a small fraction is found in concentrated ores. The uranium ores mined

The polished floor of the charge room of Dungeness power station reflecting a scrupulously clean fuel-loading machine contrasts strongly with the murky state of some coal-fired stations, and the accident rate among the operators is correspondingly less

97

Many uranium deposits, like this one (right) in Utah, are deep underground but they are not usually economically mineable unless they occur with other ores. Below: prospecting for uranium can be a simple operation as a geiger-counter will often detect its presence even through rock

at present are mainly those from which the element can be recovered for a cost less than $10 per lb. These are found in two main areas, the Witwatersrand in South Africa and Elliott Lake in Canada. Both contain the quartz-pebble conglomerate type of ore. In Canada, the average concentration of uranium oxide (0.12 per cent) is high enough for the ore to be mined for its own sake, but in South Africa the concentration is low (0.025 per cent) and economical mining is only possible because the uranium occurs with gold. Together, these deposits account for nearly 60 per cent of the known reserves of low-cost uranium. About 25 per cent more are found in sandstones, mainly in the Colorado plateau of the United States. And another 6 per cent are in vein deposits in rocks, mainly in France, but also in Spain, Portugal and Central Europe. The total amount of these reserves is about 700,000 tons.

The world demand for natural uranium will grow from the present 25,000 tons to about 50,000 tons by 1975 and to 80,000 tons by 1980. And the cumulative requirement up to 1980 will be about 600,000 tons. The known low-cost reserves will therefore be exhausted in the early 1980s. But it is probable that additional resources, at least as large as the known reserves, will be discovered and these should provide enough fuel for the world's reactors for a further 5 to 10 years. Most countries with nuclear programmes keep quantities of uranium as reserve supplies, a habit which produces a significant additional demand for the material.

It is clear that it will be necessary to utilize our uranium resources more efficiently than we do at present. This will be achieved by breeder reactors, which in fact produce more fuel than they consume, for, once these become widely used, the future requirements for uranium will begin to decrease sharply.

For many countries, an important aspect of the general problem of future nuclear fuel supplies is the availability of enriched uranium. Large quantities will be required during the 1970s to fuel gas-cooled reactors and light-water reactors. At present, the only exporters are the United States and the Soviet Union. But the capacity of the three American gas diffusion plants – at Oak Ridge, Paducah and Portsmouth – will become inadequate to meet the non-communist world's needs by about 1975. There are, therefore, powerful reasons for other countries to consider how they can best obtain their future enriched uranium supplies.

In the United Kingdom, the requirements for the AGR programme will be produced by the gas diffusion plant at Capenhurst. The military purpose for which Capenhurst was originally built was completed by 1962. And, since that time, the plant has been adapted to the requirements of the civil nuclear programme. In 1965, it was decided to modify, at a cost of £14 million, the largest stages of the plant. This will enable the output of enriched uranium to be increased considerably. But, even with the modifications, Capenhurst will only be able to supply about one-quarter of the demand in 1980. The countries of western Europe are facing similar problems – the capacity of the French gas diffusion plant at Pierrelatte will become inadequate by the mid-1970s.

An indication of the enormous electrical power demands of a gaseous diffusion plant is given by this mass of apparatus comprising only a part of the switchyard at the plant in Portsmouth, Ohio. Each of the 3 US plants, in producing fissionable material, uses as much electrical power as a large city

The Europeans have, therefore, to decide whether to rely on the United States for enriched uranium, assuming that American capacity will be increased to keep pace with the total demand, or to produce their own material. The United Kingdom, the German Federal Republic and the Netherlands, having opted for independence, have decided to collaborate in the construction of gas centrifuge plants, arguing that this is a more attractive proposition than a new European diffusion plant. Because a plant large enough to satisfy European demands will require several million centrifuges, it is necessary to obtain the economies of mass production – hence the three-nation agreement. Two experimental plants are to be built, one at Almelo in the Netherlands, and the other in the United Kingdom, probably at Capenhurst.

It is unlikely that the exploitation of nuclear power

will become restricted by a shortage of cheap fuel. But it may be a close thing, with the use of breeder reactors saving the day.

International adoption of nuclear power

Today, seventeen countries have power reactors. There are 110 of these in operation with a total capacity of about 25,000 MWe. The dates at which the first reactors in these countries began operating are: the Soviet Union (1954), The United Kingdom (1956), France (1956), the United States (1957), China (1957), the German Federal Republic (1960), Canada (1962), Belgium (1962), Italy (1962), Sweden (1963), Japan (1963), the German Democratic Republic (1966), Switzerland (1966), Spain (1968), the Netherlands (1968), India (1969) and Pakistan (1970). This list will continue to grow and, by 1975, five other countries will be operating reactors – Argentina, Formosa, Bulgaria, Czechoslovakia and Korea. By 1980, over thirty

Countries with power reactors in 1970

	total megawatts (MWe)	number of reactors
United States	11900	31
Britain	5600	30
France	2700	10
Soviet Union	1750	20
Canada	1240	4
Japan	1200	5
West Germany	990	9
Italy	600	3
Spain	590	2
Sweden	590	3
India	580	3
Switzerland	360	2
Czechoslovakia	110	1
East Germany	70	1
The Netherlands	50	1
Belgium	10	1
China	not known	—

The world distribution of nuclear power showing (colour) the countries with nuclear power by 1980

countries will be doing so. Firm plans for the installation of the first commercial nuclear plant have been announced also by Australia, Austria, Brazil, Finland, Greece, Mexico, South Africa, and Thailand. And Norway and Portugal will follow suit when they judge the conditions to be right. Other countries considering nuclear power, but less advanced in their plans, include Hungary, Iran, the Phillipines, and Turkey.

The total installed nuclear capacities up to 1975 are plotted graphically on page 103. It will be seen that the slope of the curve changes sharply after 1969, showing the increasing use of nuclear power after the establishment of its competitiveness. The estimates up to 1975 are reliable because it takes four or five years to construct and commission a power reactor, and so those which will come into operation in 1975 will have already been ordered. But, because the technology advances rapidly and economic conditions continually change, nuclear programmes are always subject to adjustment and it is not possible to forecast the total nuclear capacity in operation at times further ahead.

The figure shows that the use of nuclear power will move to a new dimension during the 1970s. By the end of the decade nearly 400,000 MWe may be in operation. And it is likely that the introduction of breeder reactors will cause a move to yet another dimension in the 1980s.

In 1975, the total installed nuclear capacity will be nearly 130,000 MWe generated by 286 power reactors. The United States will have by far the most, nearly 70,000 MWe. The United Kingdom will have over 11,000 MWe. Japan, Canada, Sweden, the German Federal Republic, the Soviet Union and France will each have between 3,000 and 10,000 MWe; Spain and Switzerland between 1,500 and 3,000 MWe; and Belgium, the German Democratic Republic, Italy and India between 1,000 and 1,500 MWe. Argentina, Bulgaria, Formosa, Czechoslovakia, Korea, the Netherlands, and Pakistan will have installed less than 1,000 MWe.

But in terms of megawatts per million of population the order is very different. Sweden will have the largest value of about 700 MWe per million persons and Switzerland will have about 400. Canada and the United States will each have between 300 and 350 and the United Kingdom and Belgium between 200 and 250. Spain and Bulgaria will have about 100. France, the German Federal Republic, the German Democratic Republic, and Japan between 70 and 90, and the remaining countries less than 50.

In most industrialized countries, nuclear power will provide between 10 and 25 per cent of the total electric power requirements by 1975, and about 50 per cent by the end of the century – in some countries (for example, Japan) it may provide as much as 80 per cent by the year 2000.

The countries now operating nuclear power reactors can be divided into two groups: those that are aiming at a large degree of self-sufficiency in nuclear energy; and those relying on experienced suppliers from other countries to install their reactors. The motives for self-sufficiency are usually related to national pride and the

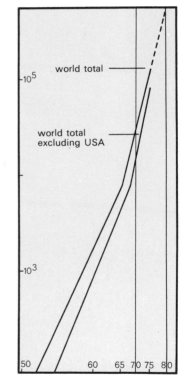

The world's total nuclear electricity-generating capacity showing the rapid increase on logarithmic scales

desire for independence, but may also include the political wish for an option to acquire nuclear weapons.

Since the 1955 Geneva Conference on the Peaceful Uses of Nuclear Energy there has been a considerable dissemination of nuclear data. Except for details of the construction and operation of uranium enrichment plants, there is much available information on all aspects of the nuclear industry and it is now possible to take many short-cuts to self-sufficiency. Nevertheless, the route must include a significant amount of research and development into some aspects of nuclear science and materials technology, and into the complex chemical processes involved. And personnel must be trained to operate the reactors and the chemical plants necessary to fabricate and reprocess fuel elements. There is much technological spin-off from a nuclear programme and a growth of subsidiary industries to exploit by-products, such as radionuclides, and to supply specialized instrumentation. But, because of the size of the financial and manpower resources involved no country would lightly embark on a comprehensive nuclear programme.

Complete self-sufficiency requires: reasonably large indigenous supplies of cheap or medium-priced uranium; the domestic production of power reactors; a plant to produce fuel elements; a uranium enrichment plant; and a reprocessing plant to extract plutonium from spent fuel elements.

The United States, the Soviet Union and China are the only self-sufficient countries. Except for adequate indigenous supplies of natural uranium obtainable for economic prices, France and the United Kingdom are self-sufficient, and India lacks only a uranium enrichment plant. Japan and the German Federal Republic have only small domestic supplies of natural uranium and no enrichment facilities. The European Economic Community would be self-sufficient if the member states pooled their nuclear facilities, although it still would not have large resources of natural uranium. Italy, Sweden and Switzerland are not aiming at self-sufficiency.

It is clearly possible to benefit from the advantages of a substantial nuclear programme without making the very large investments necessary for self-sufficiency. A relatively small cadre of personnel trained in the appropriate disciplines could, with the help of carefully selected expert consulting engineers and experienced suppliers, manage such a programme.

Nuclear power and developing countries

The number of developing countries in the list of countries operating nuclear power reactors is disappointingly small – in fact, India is the only one with a substantial nuclear programme. This is in direct contrast to the expectations of the Atoms for Peace plan but is not very surprising if the conditions for the optimum use of nuclear power are recalled. These include a high level of industrialization and a large transmission (or grid) system. The area served by a nuclear station should be able to absorb a base load of at least 500 MWe and preferably much more, and this implies an urban area of high population density. The use of nuclear power is more likely if the main indigenous fuel supply is coal, since it is less competitive with oil-fired power and still less with hydro-power. Even so, nuclear power may be introduced to reduce reliance on imported oil, for foreign exchange or strategic reasons. But there are few developing countries in which these conditions apply satisfactorily.

India is a somewhat special case. Her consumption of electricity doubles every five years, compared to ten years in most of the industrialized countries. India has four major regional electric grids, each with an installed capacity of about 3,000 MWe and annually requiring an additional capacity of about 500 MWe. Hydroelectricity provides a large fraction of India's total power but this source suffers from several disadvantages mainly arising from the seasonal character of rainfall during the monsoon. And there are many regions in which the economy is seriously affected every few years by power cuts caused by drought. Moreover, the loads in the grids are subject to large variations – the minimum can be as low as 50 per cent of the peak load. To improve this situation, India is installing reactors to provide extra power in areas affected by seasonal changes in the supply of hydroelectricity, and in areas in which users do not provide a continuous load throughout the day. Any extra power can be provided, at low cost, to encourage the growth of a number of industries. And, because Indian coal deposits are frequently far removed from the main areas of electricity consumption, the cost of fuel transportation is high. Under these conditions, nuclear electricity is economically attractive even if produced from relatively small nuclear stations. There has also been a political desire to acquire the option of producing nuclear weapons.

A main restraint to the installation of nuclear stations in the developing countries is the dependence of capital costs per unit of output on their size. This factor, far more critical for nuclear stations than for conventional ones, has caused the development of large (over 500 MWe) nuclear reactors for domestic markets in industrialized countries and a consequent lack of interest in the exploitation of the market in smaller units. But the requirements of the developing countries are often best suited by reactors of sizes between 100 and 500 MWe. This factor has been aggravated by the size of the domestic market in the developed countries which has been more than enough to keep the main manufacturers of nuclear power station equipment fully occupied. Moreover, there have been large variations in the estimates for plants of the same type and size in the United States – the major supplier – causing much uncertainty in investment costs.

Another restraint is the relatively large initial capital investment required for nuclear stations. And, because the money has usually to be found in foreign currency, which is in short supply in nearly all developing countries, and because international finance for nuclear power is, to say the least, not readily available, all nuclear plants ordered by developing countries depend on bilateral assistance in one form or another.

It is unlikely that nuclear power will make a significant contribution to the energy supply in developing countries in the 1970s. But the use of research reactors is likely to increase and there will certainly be an expansion in the applications of radionuclides for scientific, agricultural, medical and industrial purposes. Nuclear stations constructed during the decade will account for less than 8 per cent of new generating capacity and, even by 1980, the amount of nuclear electricity generated will be less than 6 per cent of the total. But the period will be used to gain experience in the use of nuclear power and to train personnel in its operation, in preparation for the time, which must inevitably come, when large amounts of nuclear energy will be used in developing countries.

The electric grids of the most industrialized of the developing countries are being enlarged to allow the absorption of the output of increasingly larger units of power. And when nuclear power is introduced in the 1980s its use will accelerate rapidly. This is about the time when fast breeder reactors are expected to become commercially viable so that much of the future energy requirements of the developing countries will be provided by these reactors.

The breeder reactor

Much development work is, at present, being done on the fast breeder reactor, the power producer of the future. These fascinating reactors differ from other types in that they produce *more* fuel than they consume.

It is possible, by a suitable design, to convert U-238 in the core of the reactor, and U-238 placed in a 'blanket' around the core, into plutonium. And 'breeding' occurs because the chain reaction proceeds with a greater neutron surplus than is possible in an ordinary reactor. The stockpile of fissile material is, therefore, steadily increased and about every ten years an amount of fuel equal to twice that put in initially is accumulated. Thus, enough fuel becomes available, not only to keep the reactor operating, but to fuel a new one of the same size. The amount of U-238 in a

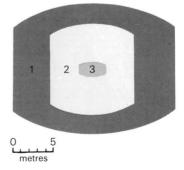

A comparison of reactor core sizes with an output of 600 MWe. 1, Magnox reactor, 2, Advanced gas-cooled reactor 3, Breeder reactor

107

The breeder reactor at Dounreay, Scotland which has demonstrated Britain's established lead in breeder reactor technology

breeder reactor is typically about fifteen times greater than the amount of plutonium, and, since about 70 per cent of this U-238 can be utilized only a small yearly replenishment of U-238 is needed. In comparison, only about 1 per cent of the uranium is utilized in light-water or graphite reactors. Fast neutrons direct from the fission process are used to induce further fission, without being slowed down by a moderator. Pu-239 is the preferred fuel for a fast reactor because more fast neutrons are available from its fission than from that of U-235. Normally, plutonium metal or oxide is mixed with natural uranium and this material is then fabricated into fuel elements. But, because most fissions are caused by fast neutrons and because the probability of causing fission with fast neutrons is only about 1/300th as great as that with very slow neutrons, a relatively large quantity of fissile material is necessary

to maintain a chain reaction. Some of the U-238 in a breeder reactor is fissioned by the fast neutrons and this process significantly contributes both to the surplus of neutrons and to the production of heat.

But there are several difficult problems to be overcome in constructing a breeder reactor, mainly arising from the large quantities of heat produced in the fuel. The core is very much smaller than those of other reactors and consequently the heat rating is an order of magnitude greater. And this gives rise to serious cooling problems. The choice of coolant is governed by the undesirability of introducing moderating materials into the core – the most suitable coolant appears to be a liquid metal, particularly liquid sodium. This material has appropriate properties at temperatures up to 500°C, which is approximately the temperature produced in the fast reactor, and it is of low

The reactor vessel during construction of the prototype breeder reactor, Dounreay. Surplus neutrons from the fission chain reaction in the core are captured in a 'breeder blanket' producing plutonium – the preferred fuel for this type of reactor

cost and readily available. But its use at these temperatures involves a completely new technology.

In a typical fast reactor, the core, the coolant, the circulating pumps and a primary heat exchanger are immersed in a tank of liquid sodium. The sodium carries the heat generated in the core into a primary heat exchanger, and another sodium circuit takes it to a secondary heat exchanger where water is heated to produce steam for the turbines.

In the United Kingdom, the first experimental breeder reactor started up at Dounreay, Scotland, in 1959 and has been supplying about 13 MWe to the County of Caithness ever since. The core of the reactor has the shape of a hexagonal prism and is provided with 367 fuel rods, 0.7 inch in diameter, made of uranium metal, enriched to 46 per cent U-235, and clad with niobium. The blanket surrounding the core contains 1,872 natural uranium fuel elements in the form of metal rods clad in stainless steel. The total weight of the uranium fuel is about 67,000 kilograms. Apart from some teething troubles, due mainly to pipes becoming blocked with sodium oxide, the reactor has worked satisfactorily.

The construction of a prototype large fast breeder reactor, to produce 250 MWe, began at Dounreay in 1967 and should be completed in 1971. And the first large commercial one in the United Kingdom will probably be commissioned in the late 1970s – it may well have a capacity of over 1,000 MWe.

The United States has been developing breeder reactors for many years. In fact, the first American reactor to generate useful, but very small (0.2 MWe), amounts of electricity was an experimental breeder at the National Reactor Testing Station at Arco, Idaho, which started up in 1951. In November 1955, the core was accidentally melted during a test in which the coolant flow was shut off and the fuel temperature permitted to rise to about 540°C. The core was re-placed with one of improved design and the reactor began operating again in January 1958. The first large American breeder reactor is the famous Enrico Fermi at Lagoona Beach, Michigan, which became opera-tional in 1963; the output is 60 MWe. The cylindrical

The Idaho Falls experimental breeder reactor where it was first shown to be possible to 'breed' more energy than was consumed. In spite of its early start in the field the US has not concentrated on breeder reactor development to the same extent as have other countries

111

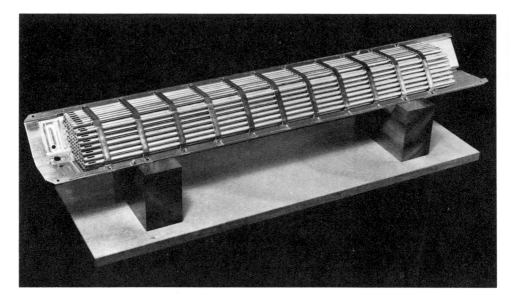

A section of a fuel element containing plutonium for the prototype breeder reactor at Dounreay

core, 2.5 feet in diameter and 2.6 feet high, contains 91 fuel elements, made of enriched uranium and clad in zirconium. And the blanket has 572 natural uranium elements in the form of rods clad in stainless steel. In 1966, an accident occurred in the Enrico Fermi when the coolant nozzles became blocked by an unsecured piece of metal. This caused serious overheating and the reactor had to be shut down for about 30 months.

The Soviet Union is devoting substantial resources to the rapid development of large breeder reactors. A number of small experimental types have been operated at Obninsk, near Moscow and, from the experience gained from these, a larger (150 MWe) reactor has been built at Shevchenko, on the Caspian Sea, partly for the desalination of sea water.

The Common Market countries are also doing a substantial amount of work on fast breeders. In 1956, the German Federal Republic, Belgium, and the Netherlands agreed to build jointly a prototype of 300 MWe. And the Republic is itself investing heavily in a major programme. In the first stage, two prototypes of 300 MWe are to be built, one steam-cooled and the other sodium-cooled. In the second stage, a 1,000-

MWe station will be built using the best of the proto-type designs. In France, a small reactor, using sodium as coolant and a plutonium-uranium-molybdenum alloy as fuel, has been in operation since 1966 at Cadarache, near Marseilles. And Japan plans to construct soon an experimental fast breeder of 100-MWe output.

Fast breeder reactors which could be built now would probably generate electricity at a cost of between 10 and 12 mills per kWh. But much lower costs, in the range of 2-3 mills per kWh, are predictable when commercial types are developed and installed in sizes of several thousand MWe.

There will be ample supplies of plutonium available for the first generation of breeder reactors. About 3 tons of plutonium are needed to fuel a breeder of 1,000-MWe capacity. In the United Kingdom, about 60 tons of plutonium will have been produced by the Magnox reactors and the AGRs by 1980, and about the same quantity will probably have been accumulated in Western Europe. In the United States, over 200 tons will be available by this time.

There are large stockpiles of uranium, depleted in U-235, accumulating from uranium enrichment plants. In the United States, 400,000 tons are likely to be available by 1980. This will be used in breeder reactors for conversion into plutonium. These reactors will there-

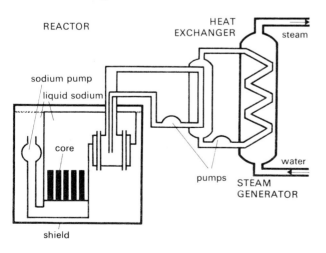

The breeder reactor using liquid sodium as the coolant. The use of this highly reactive substance required the solution of many difficult engineering problems

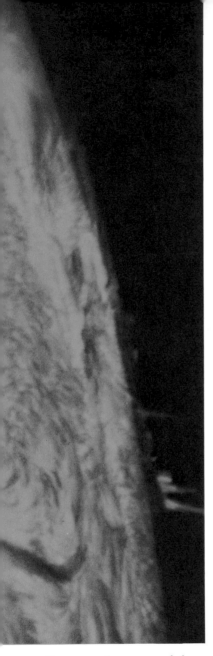

To master the process of fusion reaction would give us access to the same energy as that revealed on the surface of the sun

fore not only ensure the efficient use of natural uranium resources but will economically employ the plutonium produced in thermal reactors and the very large amounts of depleted uranium which are accumulating.

Some exploratory work has been carried out, particularly in Russia and Belgium, to investigate the possibility of using thorium in fast breeder reactors. If the naturally occurring isotope Th-232 is introduced into the reactor some of the nuclei will capture neutrons, producing atoms of the isotope Th-233, which decays to protactinium-233. Protactinium-233 decays, in turn, to U-233. The nuclei of U-233 are fissionable by slow neutrons and, therefore, it is a potentially valuable fissile material like U-235 and Pu-239.

But the use of U-233 as a fuel for reactors has not, as yet, been fully explored. Most of the thorium so far discovered is in India and this use of U-233 will be of considerable importance to the Indian nuclear programme. Thorium could be placed around the core of fast breeder reactors and the U-233 so produced used in either further breeder reactors or in ordinary reactors.

Burning the seas

Homi Bhabha, as President of the first United Nations Conference on the Peaceful Uses of Atomic Energy in 1955, said: 'I venture to predict that a method will be found for liberating fusion energy in a controlled manner within the next two decades. When that happens, the energy problems of the world will truly have been solved for ever for the fuel will be as plentiful as the heavy hydrogen (deuterium) in the oceans.' Much effort is being devoted to the problem of producing a controlled fusion, or thermonuclear, reaction motivated by the knowledge that there is enough deuterium in the oceans to supply man's energy needs for millions of years.

Fusion is the ultimate source of the energy of the sun and the stars – in which nuclei of hydrogen are constantly fusing together to form helium. In the interior of these bodies temperatures range from 5-100 million

114

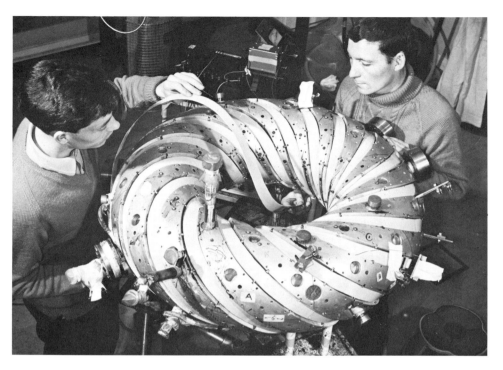

degrees, sufficient for fusion to occur – a possibility suggested as long ago as 1929, by R. d'E. Atkinson and Houtermans, in Holland. Some of the fusion energy is deposited in the interior of the star, keeping it sufficiently hot to maintain the process.

Stellar fusion reactions take place only at a slow rate. Consequently, to produce large amounts of energy in a short time even higher temperatures than those found in the stars are required. Even at the temperature of the sun it would take several million years to convert a single gram of hydrogen to helium. But no existing material can withstand temperatures higher than a few thousand degrees. And the major problem in producing a fusion reactor, which so far has not been solved, is to isolate the fuel from the walls of its containment vessel.

Controlled fusion processes can only take place in a sufficiently dense and stable plasma, a very hot gas consisting mostly of electrons and positive ions –

The doughnut shaped torus being prepared for a fusion research experiment generates an intense magnetic field and thereby confines hot plasma to a thin suspended ring as in the diagram

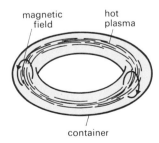

atoms which have lost one or more electrons – in nearly equal concentrations. In current laboratory experiments, the use of magnetic fields to isolate the plasma from the container walls is being studied.

Several countries are performing fusion research, including the United Kingdom, the Soviet Union, the United States, France and the German Federal Republic. The six Common Market countries are co-ordinating their research through Euratom. And, since 1958, there has been a full exchange of information between all these countries.

At present, the Soviet Union is probably the farthest advanced. In recent experiments, Soviet scientists have produced all the requirements for the fusion process simultaneously and for an unprecedented length of time. The Soviets use a toroidal system, shaped like a doughnut, which confines the plasma by a magnetic field in a chamber having an internal diameter of about 10 cms. The machine itself, called Tokamak-3, is about 2 metres across. A hot plasma has been confined for times exceeding one-twentieth of a second at a temperature of 10 million degrees centigrade and a density of 2×10^{14} ions per cubic centimetre. This is regarded as a very significant advance in demonstrating the feasibility of the practical utilization of fusion.

Accurate measurements of the temperature and density of the hot gas were carried out by a British team of scientists, from the Culham Research Laboratory, using an advanced laser technique developed by them. The complex apparatus weighing 5 tons, was transported to Moscow to measure the parameters of the plasma more accurately than any other known method. This collaboration, the first in which British and Soviet scientists have worked together in an experimental nuclear energy research project, is an example of the growing international co-operation in fusion research.

It is generally reckoned that the conditions necessary for the operation of a fusion reactor are a temperature of about 100 million degrees centigrade, and a combination of density and confinement time that yields

a product of about 10^{14}. Thus, a fusion reaction might operate with a density of 10^{14} ions per cubic centimetre provided that the confinement time was one second. But, for the Soviet confinement time, a density of 2×10^{15} ions per cubic centimetre would be required. The next step is to increase both the temperature and the density-time product another 10 times. This will be a formidable task even though, in the past decade, fusion scientists have achieved increases in temperature of 200 times and in the density-time product of about 50,000 times. The Soviets intend to build a larger Tokamak in an attempt to achieve longer confinement times.

It is very difficult to predict when fusion reactors will be commercially developed – past hopes have been proved over-optimistic. But it is unlikely to be before the year 2000. The USAEC have forecast that the feasibility of controlled fusion will be demonstrated by 1978 but, because of the recent Russian progress, this could occur earlier than this date – possibly in the new Tokamak.

When a fusion reactor is eventually built it will probably utilize a plasma of deuterium and tritium contained in a toroid, possibly about 75 metres across.

One of the devices used in experiments aimed at achieving controlled thermonuclear fusion which has proved successful in confining hot plasma

117

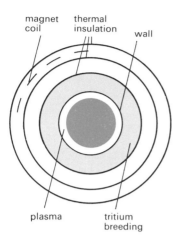

The elements of a possible fusion reactor in vertical section

The plasma will be surrounded by a 'blanket' through which a coolant will flow to remove the fusion energy from the reactor. The heat produced will then be used to generate electricity as in a conventional power station. And, it is likely that the blanket will be used for a second purpose – to breed tritium from lithium, incorporated in the blanket, by neutron bombardment. Sufficient tritium could be produced in this way to replace that consumed by the fusion process in the plasma. Apart from the initial fuel, the reactor will therefore be fuelled by deuterium and lithium. The wall separating the coolant from the plasma could be constructed from, for example, molybdenum, a material which has good high temperature properties and which does not capture neutrons appreciably.

In one concept of a commercial fusion reactor, fissionable material is also used in the blanket. If U-238, for example, was incorporated, the fast neutrons emitted by fusion reactions would produce fission, and a large fraction of the energy would then come from this process. The reactor would require a smaller and therefore cheaper, magnet, but would have increased radioactivity hazards arising from the fission products in the blanket.

Speculative predictions have been made of the construction costs of fusion reactors and of their electricity generation costs. They are likely to be installed at a time when very large commercial power stations are required. The capital cost of a 2,000-MWe reactor will probably be about £60,000 or £70,000 per MWe and the generation costs will be about 2.5 mills per kWh. These costs are comparable with those forecast for large breeder reactors in the 1990s.

A major advantage of the fusion reactor will be the lack of build-up of radioactive waste products. In fact, the main fusion product will be non-radioactive helium. And the fuel supplies for fusion reactors will present little problem. Deuterium is cheap and reserves are virtually inexhaustible. Present supplies of tritium are produced by the neutron irradiation of lithium in reactors. The estimated cost of large-scale production is about £1,000 per gram, but the tritium bred in fusion reactors will cost ten times less than this. A 2,000-MWe reactor will probably require a daily fuel intake of only about 2 kilograms. And, after the development of fusion reactors, uranium will suffer the normal fate of major fuels and be eventually superseded, by deuterium.

In both fission and fusion, only a very small fraction of the available mass is converted into energy. If the conversion were total, an enormous amount of energy would be produced – one pound of matter would be equivalent to 1,500,000 tons of coal. And it is intriguing to speculate that, in the distant future, man's energy requirements may be met by the controlled release of all of the energy of the atom by its total destruction. But, for the forseeable future, the fusion reactor will be the end of the long and winding road taken by nuclear power technology. We must now turn to the other important applications of nuclear energy.

In 1957 the founders of the International Atomic Energy Agency stated that the uses of radionuclides 'promised almost infinite possibilities for the advancement of knowledge, for the improvement of industrial processes and for economic and social progress.' The multitude of successful ways in which radionuclides are now used in medicine, agriculture and industry proves the validity of this early assessment; and it has become clear that the contribution made by radionuclides to our welfare is equal in importance to the contribution made by nuclear power. These nuclear tools can be used almost everywhere – to increase food supplies and improve the quality of food, to fight disease and improve health standards, to increase industrial output and reduce the cost of goods, and to develop supplies of water. In addition, they have been employed with considerable success in pure scientific research to make many fundamental discoveries – particularly in chemistry and biology – which would not have been possible by conventional methods. And the developing countries have already derived great benefit from radionuclide techniques, whereas nuclear power has done little so far to narrow the gap between them and the developed countries.

Medicine

Radionuclide techniques are widely used in medical practice and many have become so firmly established

The use of radionuclides in many fields has opened new possibilities for science. Here scientists are shielded by 7 inches of lead as they prepare labelled chemicals at the Radiochemical Centre, Amersham

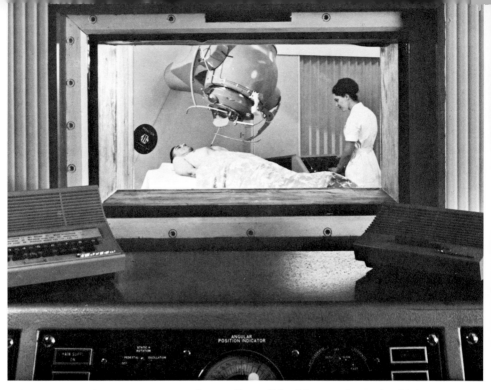

A large quantity of cobalt-60 is the source of radiation in this cancer therapy unit where the patient can be viewed through a four foot thick lead-glass window during the course of treatment

Opposite: a cancer therapy unit at Churchill Hospital, Oxford

that radionuclide laboratories exist in most major hospitals throughout the world. Radionuclides are used in the diagnosis and treatment of certain diseases, in the study of biological processes within the human body and of the effects of specific diseases, and in the study of disease-carrying organisms. Although much progress has been made in medical applications, the number of different radionuclides at present used is smaller than might have been hoped. Over one thousand different species of radionuclides have been discovered and there are now nearly two hundred of these offered commercially. In addition, a very large number of chemical and physical variants is available. But only about fifty radionuclides have been considered for medical use and, of these, only about one-half are used in established routine applications.

The effective use of radionuclides for therapeutic (treatment) purposes depends on the ability of the radiation (alpha, beta or gamma) they emit to produce ionization of the atoms of the matter through which it

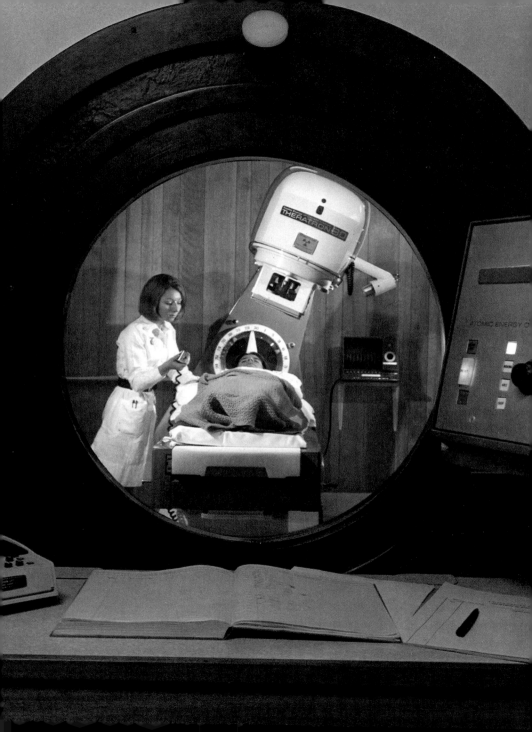

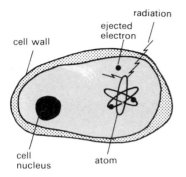

Above: radiation, by detaching electrons from molecules, can make highly reactive ions which damage or kill the cell

Below: cancerous cells – the target of radiation treatment. The smaller cells are red and white blood cells

passes. An atom is ionized if one of the electrons moving in orbits around the nucleus is removed from the atom. And if an alpha particle, a beta particle or a gamma ray collides with an orbital electron, enough energy may be transferred to the electron to ionize the atom. Ionization occurring within a cell of the body often results in the death of the cell, whether it is healthy or diseased.

The most usual targets of radiation therapy are cancer cells, which grow and divide more rapidly than the normal cells of the body to form a growth or tumour. Radionuclides are employed for therapeutic purposes in a variety of ways. Cobalt-60 and caesium-137, which have long half-lives of 5.2 and 27 years respectively, are used, in units of large quantity, as external sources of gamma radiation in a manner similar to an X-ray machine.

In a few cases, it is possible to administer to the patient a radionuclide which becomes concentrated in a specific organ by the normal metabolic processes. And if this organ contains a tumour, for example, the cancerous cells will eventually be destroyed by local irradiation. It is inevitable that some healthy cells around the site of the cancer will be killed off also, but the body is able to cope with this damage provided it is restricted in area. This is the ideal method of radionuclide therapy and it is a great disappointment of nuclear medicine that, in practice, it can be achieved in only a small number of cases, such as the treatment of diseases of the thyroid gland by the administration of radioactive iodine, the treatment of polycythaemia vera and leukaemia by radioactive phosphorus, and the treatment of leukaemia by radioactive gold.

Polycythaemia vera is unique in that it is, so far, the only disease for which treatment with radionuclides is the only effective cure. It is an uncommon disease of the blood which results in the formation of too many red blood cells; its cause is not known. The white cells are also overproduced, but to a lesser extent than the red cells. The first sympton is a reddening of the skin, and possibly there are headaches. This is

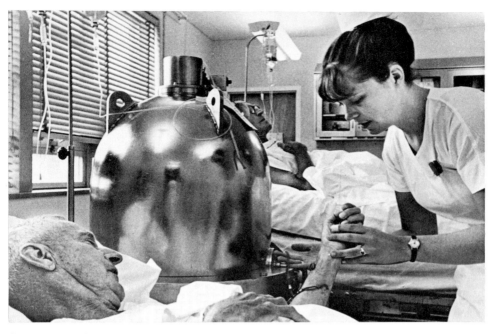

followed by thickening, and possibly clotting, of the blood, which makes the patient abnormally prone to heart attacks, blackouts, strokes, ulcers, and varicose veins. The disease has been treated by bleeding, by drugs such as phenylhydrazine, and by X-radiation, but each method has serious disadvantages. Bleeding is unpleasant and burdensome for both patient and doctor, phenylhydrazine dosage is difficult to regulate and does not control the production of the white cells and the X-radiation of bone and spleen exposes the patient to excessive radiation.

But treatment with phosphorus-32 is simple and effective. A small quantity of the radionuclide is injected into the blood stream and eventually becomes uniformly distributed in the blood and bone marrow. The irradiation of the bone marrow by the beta-particles emitted by the phosphorus slows up the production of the red and white cells and, after a period of about a month, the red-cell count begins to fall. The delay is due to the long life of the red cells; the white cells, because of their shorter life, become less numer-

Above: treatment of leukaemia by radiation

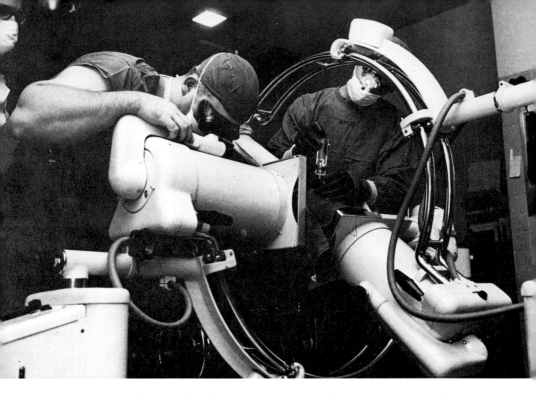

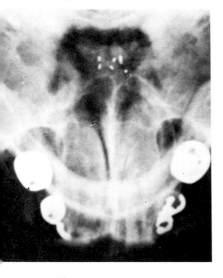

Radionuclide pellets implanted (above) with X-ray guidance image appear (below) as 6 white dots in the pituitary at the base of the skull

ous quite soon after the injection. The effect of the radiation is indirect, since it is the production of the blood cells – by the bone marrow – which is affected rather than the cells themselves.

The treatment of polycythaemia with radioactive phosphorus was first attempted in 1936 by Lawrence, an American doctor, and many patients have since been treated by the method. It appears that their life expectancy after treatment is about normal and that no serious complications arise. The earlier methods of treatment gave an average length of life of only six or seven years after diagnosis.

Other methods of delivering radiation to cancerous tissue, by the local application of radionuclides, have been successfully developed. Radioactive sources can, for example, be actually implanted into tumours – by surgery, if necessary – in many sites of the body, and a variety of sources are available for this purpose. Needles, wires, cylinders, sintered rods and ceramic beads, all containing radionuclides, are frequently used. And radioactive solutions, confined in balloons, are used to irradiate tumours in cavities within the body, such as the urinary bladder. Solid radioactive sources – for example, small spheres impregnated with a radionuclide – are also available for intracavitary treatment. Thus, in the treatment of tumours of the oesophagus, a number of these spheres are threaded on a string and swallowed by the patient. After the required irradiation at the site of the tumour the spheres are removed by withdrawing the string.

A seed of yttrium-90 is introduced through the nasal cavity to the pituitary to suppress its activity in a case of malignant cancer

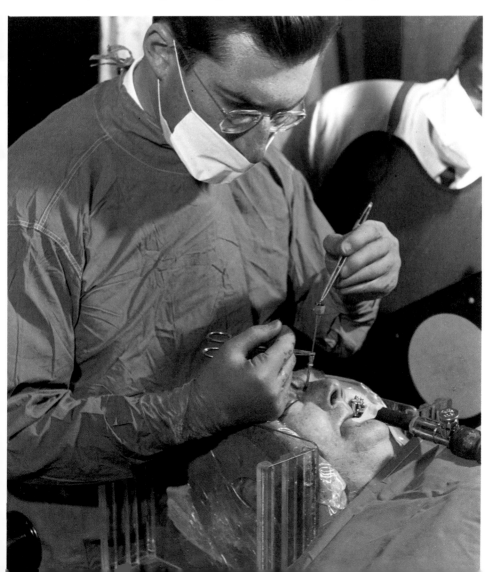

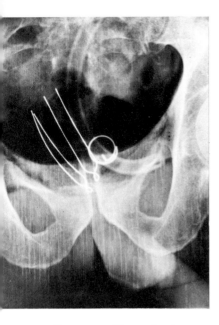

Above: an X-ray showing two 'hair pins' containing the radionuclide tantalum-1827 implanted into the bladder to irradiate a tumour

Compensation for the unattractive prospect of entering a whole body counter might be the timely detection of nutritional deficiencies by an assessment of the child's natural radioactivity

The main diagnostic uses of radionuclides can be subdivided into: studies in which body-space volumes are measured – for example: the measurement of the volume of red blood cells or circulating plasma; studies of the functional state of body organs and tissues, such as the measurement of kidney function or the survival times of red blood cells; studies of the absorption of important substances like iron, calcium or vitamin B-12; and the localization of tumours. In one hospital alone – the Johns Hopkins in New York – over 100 patients daily are given radionuclides for diagnostic purposes. Most diagnostic studies are performed using very small quantities, or tracers, of radiopharmaceuticals – chemical compounds 'labelled' with radionuclides. A radiopharmaceutical is produced by replacing a stable atom of the molecule by a radioactive atom of the same element. And, in a radioactive tracer technique the fate of the substance of interest in the system under investigation is followed by introducing a tracer quantity of a suitable radiopharmaceutical and following what happens to it by measuring the radiation emitted. This is well illustrated by the diagnostic use of radioactive iodine for testing

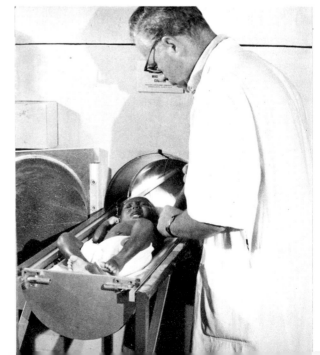

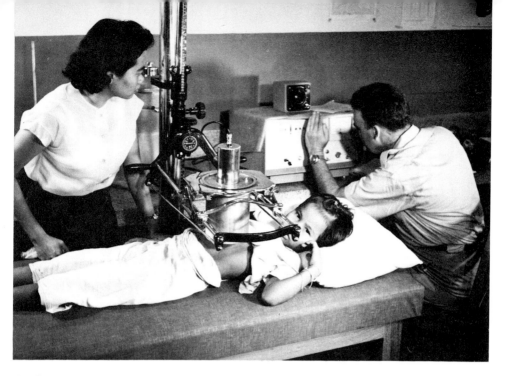

the functions of the thyroid gland – the most widely used clinical application of radionuclides. The accuracy of diagnosis of thyroid disease is considerably increased when the results of this radionuclide test are added to the clinical assessment of the patient.

The thyroid gland removes iodine from the blood to synthesize its principal secretions, the hormones thyroxine and triiodothyronine. The use of radioactive iodine enables an estimate to be made of the gland's activity. The two most frequently encountered thyroid disorders are hyperthyroidism and hypothyroidism in which the secretion rate of the gland is increased and decreased respectively. But a direct measurement of the secretion rate of the hormones is not a practicable routine diagnostic procedure. In the radionuclide test, the patient is given a drink containing a small quantity of radioactive iodine and the amount of radioactivity taken up by the thyroid gland is measured by detectors, suitably placed outside the body. In hyperthyroidism the proportion of the administered radioactive iodine found in the thyroid is

Spleen function is monitored by tracing radionuclide labelled blood. Below: radioactive food used in a study of venom

greater than that found in normal subjects, while in hypothyroidism it is less.

A clinical application for which there is no satisfactory alternative is the use of radioactive cobalt for the measurement of the metabolic behaviour of vitamin B-12. This is of particular value in the diagnosis of pernicious anaemia, a condition caused by failure to absorb this vitamin from food. Vitamin B-12 contains a cobalt atom in its molecular structure and it is, therefore, possible to label it with radioactive cobalt. If the patient is given a small oral dose of the labelled vitamin, the amount absorbed by the body can be measured and abnormal absorption detected; this is most commonly achieved by measuring the amount of radioactivity excreted in a 24-hour collection of urine.

The measurement of the distribution of elements within the body is the most varied diagnostic application of radionuclides. Many different radionuclides have been used for this purpose and their distributions in most organs are investigated. Techniques have been developed to provide an actual 'picture' of various

Above: in suspected hyperthyroidism uptake of imbibed iodine-131 is measured and compared against a graph like the one below

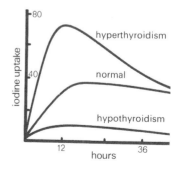

Opposite: production of radiopharmaceuticals

131

A scanning scintillation counter measures the uptake of radioactive-iodine revealing the condition of the thyroid. The left-hand one is normal, the middle one has a nodule causing localized overactivity and the right-hand one is underactive

organs of the body so that tumours, cysts and other lesions can be located. This science (or art), called scintigraphy, provides a unique method of obtaining certain types of medical information. Medical scintigraphy is based on the capability of certain organs to accumulate, either temporarily or permanently, specific radioactive substances after they have been given to a patient by mouth or by injection. Subsequently, the distribution of radioactivity is 'mapped' using an instrument called a scanner. The pattern obtained allows some useful conclusions to be drawn concerning the size of the organ and its normal or abnormal position in the body. A concentration of radioactivity in a place where there is normally none, or the absence of radioactivity where there should be some, may indicate the presence of a tumour, or other lesion, which might not be detected otherwise.

A scanner consists of a radiation detector which moves over the patient and is made to 'scan' automatically, line by line, across the suspected areas of disease. And different levels of radioactivity are often indicated by dot patterns on a sheet of paper. The dots are closer together when the detector is over a region of high activity and are more widely separated when the detector is over regions of low radioactivity. Alternatively, colours are used to assist the doctor in interpreting the results. Thus, red dots may indicate the presence of high levels of radioactivity and different coloured dots the lesser levels.

Recently, a stationary detector device, called a

gamma camera, has been developed with the advantage that it enables an almost instantaneous 'picture' of radioactive regions within the body to be taken. With this instrument, it is possible not only to record the distribution of the radioactivity fixed in a particular organ but to follow its movements through a number of organs such as the kidneys, heart and lungs.

Scintigraphic methods are particularly useful for obtaining information about such organs as the brain or the thyroid gland, which ordinary diagnostic radiology is often incapable of obtaining. If it is suspected that a patient is suffering from a brain tumour, for example, he may well be sent to a hospital which has a

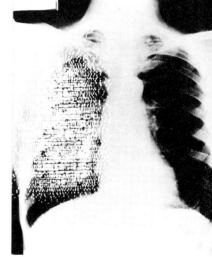

Above: a lung scan superimposed on an X-ray showing how an administered radionuclide (gold-198) is distributed, in this case, throughout the pleural cavity

Left: radiation detectors recording on a graph above the patient's head the concentration of radioactive iodine-131 labelled hippurate in a check on kidney function

133

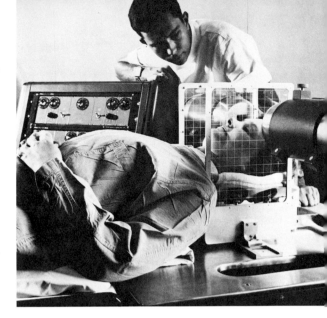

Right: the accumulation of administered radioactive copper-64 and arsenic-74 detected by a scanning instrument indicates a brain tumour

An X-ray superimposed on a brain scan taken after the administration of radionuclides shows the position of a tumour

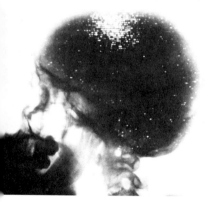

department of scintigraphy. Here, he will be given an intravenous injection of a small quantity of a suitable radionuclide, probably one of the radioactive isotopes of arsenic. About one hour later, the scanning of his brain area will commence. The patient is usually scanned while lying on his back and then a second scan is often taken while he lies on his side. The complete operation will take about one-and-a-half hours. And the scan can be repeated on the second day to observe transient processes. Because arsenic tends to localize in brain tumour tissue, the presence of a tumour will be indicated on the scan by a region of high radioactivity. This process is assisted by a unique property of the brain, the so-called blood-brain barrier. Apart from a few chemicals, most tracer materials are excluded from the brain under normal circumstances. But if the brain is disturbed by, for example, the growth of a tumour, this barrier mechanism breaks down and previously excluded chemicals, like arsenic, can enter the brain tissue.

Certain radionuclide applications are of great clinical value because of their unique speed and simplicity. For example, a protein called human serum albumin, labelled with radioactive iodine, can be used to

measure total blood volume which is often of considerable diagnostic importance. The patient is given
an injection of the labelled albumin and a small sample
of blood is taken shortly afterwards. This is put into an
instrument which automatically computes the volume
of blood in litres. Another important application is the
use of iodohippurate, also labelled with radioactive
iodine, for tests of kidney function. This substance is
extracted from the blood, by filtration, in one passage
through the kidneys. The test consists of recording the
accumulation and disappearance of radioactivity, using
detectors placed over the kidney area, after the intravenous injection of a small quantity of the labelled
compound. This technique, called renography, is a
convenient and safe procedure in the early investigation of possible kidney disorders. The same compound
can be used for measuring residual bladder urine without the hazards of catheterization, a particularly
valuable technique in diabetic patients.

The use of radioactive gases to investigate lung
function is a noteworthy recent development. One
example of this is the measurement of blood flow in the
lungs. A saline solution of the radioactive gas xenon-
133 is injected intravenously into the patient. And when
it reaches the lungs most of the gas passes out of the
blood into the air in the lungs because of its low solu-
bility. The xenon is then removed at a very slow rate
from the lung by the blood flow. The patient holds his
breath for a few seconds during which time the distribution of radioactivity in the lung is measured by
external detectors. This will show whether the blood
supply and the blood-gas exchange is uniform throughout the organ.

The applications discussed so far involve giving
radioactive material to the patient and making measurements on him in hospital. But there exists a group
of important techniques in which no radioactive
material is given to the patient and in which the necessary procedures are performed in the laboratory on
samples of blood, urine or body tissue. The advantage
is that this can be carried out, on samples collected

from a wide area, at some central laboratory. In a typical group of procedures, small samples of blood or of body tissue, obtained by minor surgery, are taken and incubated with a compound labelled with a radionuclide. The way in which the substance is metabolized by the tissue is observed to provide information in the diagnosis of cancer and other conditions.

In another interesting group of procedures, radionuclides are used to test the immunity of populations to infectious diseases, such as cholera and plague. The immunity of an individual to an infection often depends on the existence in the blood of certain antibodies, substances which can combine with the infecting organisms, rendering them harmless. It is often necessary to know the extent to which a population, threatened by an outbreak of an infectious disease, is immune so that mass immunization can be planned, if and when necessary. For this purpose, a method of examining blood for antibodies against specific bacteria is being developed using bacteria labelled with a radionuclide. Only a few drops of blood are required and the test can be easily carried out on a large number of subjects when an outbreak of a disease occurs.

Agriculture

Radionuclides have had the greatest impact in agriculture in the breeding of new varieties of plants and in obtaining the largest possible crop from the minimum use of fertilizers. Plant crops, particularly cereal grains, are the basic source of protein for most of the world's population, especially in the developing countries. But the natural protein content of these crops is low. Increasing their quantity and nutritional quality is potentially the best way of combating world-wide protein malnutrition. And this has been achieved by exposing seeds to the radiations emitted by radionuclides to increase vastly the rate of mutations in the seeds. Radionuclides are also used to select only those mutants of the highest nutritional value.

Mutations are changes in the hereditary nature of living things and one of the main advantages of muta-

tion breeding is the ability to improve one characteristic of a variety without altering to any extent the otherwise desirable characteristics. For example, the weak stem of an otherwise well-adapted variety of plant can be shortened and strengthened, resulting in higher crop yield. Mutations occur spontaneously, but at a relatively slow rate and only a very small fraction of these turn out to be beneficial. But it is possible, by applying radiation, to increase the rate from between 10,000 and 100,000 times so that every plant in an irradiated population contains at least one mutation. The success of this technique has already been dramatic. Dozens of new varieties of plants have been evolved, including certain types of rice, wheat, barley, oats, peanuts and soyabeans – all having high protein content, increased crop yield, and improved resistance to disease and weather. Millions of acres of these superior crops are now being grown throughout the

With growing demands for crop improvements radiation induced mutations in plant breeding offer enormous possibilities. In a special greenhouse plants are irradiated by a radionuclide which can be raised from a central, shielded pit to administer regulated amounts

Radioactive tracers help determine the optimum use of fertilizers on rice (right) and show the distribution in a leaf (below) of fertilizers placed in the soil

world. The potential of the method is indicated by reports of a doubling of the protein content of rice. And the highest-yielding rice variety in Japan, called Reimei, is a radiation-induced mutant.

A significant reduction in the growing period of rice has also been achieved by the use of mutants. In Hungary, for example, a shortening by three weeks has been obtained which could have far-reaching consequences on the extent of rice growing in Europe. And a new variety of castor oil seeds, evolved in India, matures in 120 days, compared with 270 days for other varieties. The soil is released for an extra 150 days for other crops such as rice.

A commercially valuable aspect of the technique is its introduction into the breeding of ornamental plants. One example among many is the beautiful Desi rose, produced in East Germany, which has dark red stripes on yellow petals.

To achieve the best performance of a plant it is often necessary to add nutrients to the soil in the form of fertilizers. But these are expensive and often scarce. It is therefore important that they are supplied at the right time and placed correctly in the soil, in the optimum chemical form, for the most efficient uptake by the plant. And radionuclides provide the only direct method of studying the fate of fertilizers and their movement into plants. For example, by labelling a phosphorus fertilizer with radioactive phosphorus it is possible to determine the place in the soil from which a plant is best able to take up this essential element.

Tree crops, such as coffee, coconuts, citrus fruits and cocoa, are of major importance to the economies of many developing countries. And it is essential that the minimum amount of fertilizer should be used on them to maximum benefit. A relatively simple tracer technique, with a radioactive phosphorus solution, is used to study the uptake of fertilizers by these tree crops. The solution is injected at various distances away from the tree-bole and at various depths. The uptake of

Selective gene manipulation by irradiation provides a method for accelerating the breeding process. A mutant of maize is made shorter to give a higher yield; a variagated rose mutant – the Desi rose; and a black mouse whose skin has been exposed in bands to radiation destroying the pigment forming cells to leave colourless hair

Above: studying metabolic pathways in protein production with labelled sulphur. Below: Mediterranean fruit flies irradiated to sterilize them

radioactivity by the tree, determined by examining samples of the foliage at regular intervals, is a measure of the efficiency of the fertilizer. Results from Ceylon indicate what this technique can achieve. Fertilizer was commonly applied in rings around coconut palms about a metre from the boles, whereas it was discovered that the greatest uptake occurred between 50 and 100 cm from the trees. In Kenya it was found that the roots of coffee bushes are almost inactive in the January-February dry season but in April-May the majority of active roots are concentrated close to the stem. In the past, fertilizer was applied some distance away, causing a serious loss of efficiency.

Radionuclides play an important role in domestic animal husbandry, even though radiation has not so far been used for breeding livestock. One of the most serious problems associated with livestock production is the loss caused by helminthic diseases (internal parasites). And parasite control is exceedingly difficult. Vaccines like those used against bacteria or viruses cannot be produced, and the only known method of obtaining helminthic vaccines is to weaken larvae (of, for example, sheep lung-worm) by exposing them to radiation.

Radionuclides are used in animal nutrition to study the metabolism of particular nutrients in the animal and their deposition in various tissues and products. For example, the pathways of nitrogen and sulphur in protein synthesis have been determined in this way.

The largest cause of food loss is insect attack on both livestock and crop plants at all stages of growth. The tsetse fly prevents the use of three million square miles across the equatorial belt of Africa for raising livestock and it is also a carrier of trypanosomiasis (sleeping sickness) which is often fatal in man. For over sixty years, attempts have been made to eradicate this pest by clearing the bush, spraying insecticides, and driving out game animals. Some success has been achieved in controlling human mortality from sleeping sickness. In 1965, only 3,000 new cases were recorded with less than 10 per cent mortality, whereas 70 years ago about

200,000 persons died annually of the disease in Uganda alone. But the disease is now on the increase again for, like all insects, the tsetse fly can develop immunity to insecticides. Accordingly, the use of radiation to sterilize vast numbers of these insects is now being attempted. If this is successful, sterile flies will be released in infected areas to compete with the existing population during the mating season. And subsequent releases of overwhelming numbers of such flies should, within a few generations, eradicate the species.

This so-called sterile-male technique has already been successfully used to eradicate the screw worm, another serious animal pest, from the United States, and is also being used to control and eradicate the

Below: screw-worm pupae sterilized by radiation to be released, 5 million a week over 50,000 square miles of the US, to compete with the natural population and diminish their numbers by introducing infertility. In this method there is no danger of contaminating or poisoning the environment as with insecticides

Mediterranean fruit fly, the rice stem-borer, and the olive fly. The method has several attractions: it affects only one species of insect and does not harm other wildlife; it is the only method which can totally eradicate an insect species from a given area; and no harmful chemical residues are left.

In all parts of the world, vast amounts of chemical insecticides are being used by farmers to spray crops. The residues left by these chemicals are a growing hazard to man and wildlife. And, again, radionuclides provide the best way of studying the presence of these residues in the food chain and of detecting any small but harmful quantities of them.

Industry

Radionuclide devices are standard equipment in modern industry. More than one thousand applications save hundreds of millions of pounds each year in running costs, and assist the industrialist to make new and improved products and to increase the speed and efficiency of his chemical and engineering processes.

One of the most familiar applications is the production of self-luminous compounds for instrument dial markings and pointers by mixing a radionuclide with zinc sulphide – the radiations emitted by the radionuclide are able to stimulate the luminescent process in luminous materials like zinc sulphide. Another useful property is that they are able to render photographic emulsions developable. An object placed between a source of gamma-radiation and a photographic film will cast a shadow because it absorbs some of the radiation. And when the film is developed it will show a shadow picture, or radiograph, of the object, similar to a normal X-ray picture. Radiography is used extensively for the detection of flaws in a wide variety of objects such as castings, forgings, welds and manufactured parts. It is also used to examine sections of large structures, such as struts in bridges. In many cases, radionuclide sources offer distinct advantages over X-ray machines. They are much more mobile, can be pushed into narrow tubes and confined spaces and are cheaper. And large thicknesses – up to about 15 cm of steel – can be radiographed with them.

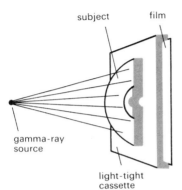

Above: the principle of radiography by which the image of a solid object is projected like a shadow by a beam of gamma rays onto a film

Below: a thickness gauge compares the penetration of rays through a sample with that through a standard of the same material

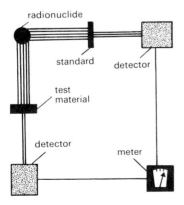

In numerous industrial processes, insulating materials may become electrically charged by passage over rollers or by friction against parts of machines. This may give workers electric shocks or produce sparks which could ignite inflammable vapours in the air. Moreover, static electricity can be very troublesome in causing cut lengths of insulating material to wrap and jam around rollers or refuse to stack properly. These problems are met in such activities as the printing of thin plastic wrapping materials; in the textile industry, through the tangling of warps or the deposition of dust on delicate fabrics; and in the pharmaceutical industry, in which certain powders can become so highly charged electrically that they stick to the delivery chute rather than fall into the containers in which they are being packed. Metal foils containing radionuclides are often used to ionize the air in the immediate vicinity to such an extent that the accumulated static charges will leak away.

Radioactive thickness gauges are extensively used to measure and control automatically the thicknesses of materials like plastic sheeting, linoleum, paper, and thin metal foils as they emerge from production machines. The gauges are based on the principle that as beta particles pass through a sheet of material they are slowed down and stopped, so that the number emitted from a radioactive source which can penetrate the sheet to a detector on the other side is a measure of the sheet's thickness. And the use of the gauges reduces wastage of material and improves the precision of production. If radionuclides which emit gamma rays are used, sheet steel up to 6 or 8 cm thick can be measured and controlled since gamma radiation can penetrate much greater thicknesses than can beta particles. This versatile technique can be adapted to measure directly the thickness of liquid or thin plastic films on rollers and also the thickness of one material coating another. By a suitable choice of radionuclides, a wide range of thicknesses can be covered – from films less than a thousandth of an inch thick to sheet metal several inches thick.

Radiography is used (left) to reveal the structure of an ancient bronze Buddha by exposing special film, suspended behind the statue, to a cobalt-60 source (on the pole) in front; and (above) to provide an image of the structure of a pipe joint on a film wrapped under the pipe

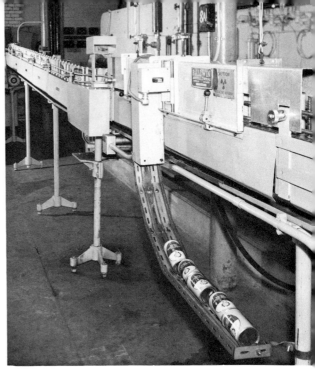

Above: a snow-density gauge.
Right: another industrial application of radionuclides. Unfilled cans are rejected as they pass the level-detector whose principle (below) is simple: if radiation is detected through the can it means the liquid level is not high enough to catch it

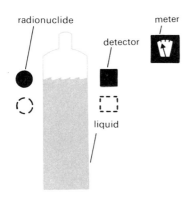

One variety of radioactive gauge, small enough to be lowered down a narrow borehole at the end of a cable, measures the density of underground strata and is used to discover deep coal seams. A similar device is employed to check the state of the foundations under aircraft runways. Mechanical coal cutters often have gauges attached to the cutting head to steer them accurately along the lower layer of a coal seam and to prevent them cutting uselessly into rock.

The radiations from radionuclides are employed in less spectacular ways – for example, to detect and sort empty or partially filled packets of various products as they pass out of filling machines or along conveyor belts. Or they are used to measure and control externally the levels of liquids in tanks or other containers, like fire extinguishers.

Radioactive tracers have various important industrial uses. The chemical industry, in particular, uses them for rapid analyses and to avoid the need for difficult chemical separations. Very small amounts of impurities can often be measured by putting samples inside a research reactor to make the samples and the impurities radioactive. When removed from the reactor, the impurity will emit its own characteristic radiation and this allows the quantity present to be measured. For most elements, amounts of less than one five-hundredth of a gram can be estimated in this way.

Radioactive tracers can sometimes be used to locate leaks in pipes and plants. To find large leaks in pipes, the tracer can be fixed in a rubber ball which fits the pipe snugly and is carried by the current as far as the leak. Here it will lodge and its position can then be located by a detector moved over the pipe. Leaks in buried pipes can be found even through large thicknesses of soil, brick or concrete. To find smaller leaks a 'slug' of radioactive solution can be passed through the pipe and, when this reaches the leak, some of the solution will pass out of the pipe into the surroundings. A detector, pulled through the pipe by a cable, will indicate the position of this radioactivity and thus the location of the leak.

Below left: locating a leak in a water-main by determining the position at which a tracer has entered the surroundings of the pipe. Below: with radiation treatment traces of ammunition on a gunman's hand can be closely identified with those on a bullet

In the motor-car industry, radioactive tracers are used to measure the rate of wear of components under different conditions of engine running. For example, if a radioactive piston is incorporated into a test engine, minute amounts of worn-off debris can be detected on the cylinder wall or in the oil sump. The properties of engine oils have been examined in a similar fashion.

In another application the rate of wear of brick linings in blast furnaces is measured by incorporating radioactive pellets into the linings at points where they can be checked periodically from the outside. When one of the pellets can no longer be detected, samples of iron taken from each cast removed from the furnace are examined for radioactivity to find when the pellet came out of the lining. This information provides a measure of the rate at which the furnace linings are being worn away.

Tracers with very short half-lives, such as sodium-24, are used to check the intimacy and uniformity of the mixing of materials in the manufacture of gramophone records, animal foods, and even confectionery, by measuring the relative amounts of the radioactive substance in samples after mixing. Because of the short life-time of the tracer, the radioactivity will have practically disappeared within a few days after the test. And these radionuclides have been successfully attached to suitably-sized carbon granules to simulate yeasts and bacteria so that the efficiency of filters dealing with these organisms in industrial processes from which it is necessary to exclude them can be determined.

The use of very large quantities of radiation for industrial processing has become the basis of a new and fascinating technology. Each year, raw materials worth several hundred million dollars are converted by radiation into newer and better products. The merits of the radionuclides cobalt-60 and caesium-137 make them particularly suitable sources of radiation for this purpose. Wood or fibres, impregnated with plastic are irradiated to produce new materials for building and the manufacture of some tools. By this process, extremely hard wood can be created from soft wood and the product is very resistant to wear, decay and insects. Also it will not burn easily, it is long-lasting, will retain its shape and is easy to work on with machines. And it can be made in a variety of attractive colours. A similar impregnation and irradiation technique has produced stronger concrete for areas where extreme service conditions are encountered.

Impregnated fibres such as bagasse, made from the residue of sugar-cane, bamboo and jute, can be adapted for building outside walls – a development of importance to countries in South East Asia and the Far East where large quantities of sugar-cane are grown and where there is a shortage of hard wood.

There are already four wood-plastic manufacturers in the United States, one in France and one in England, producing flooring, cutlery handles and other pro-

Delighted by the prospects of longevity this carving by a Thai craftsman has been impregnated with plastic and subjected to radiation – a technique which makes wood into a hard, durable substance resistant to fire, insects and damp

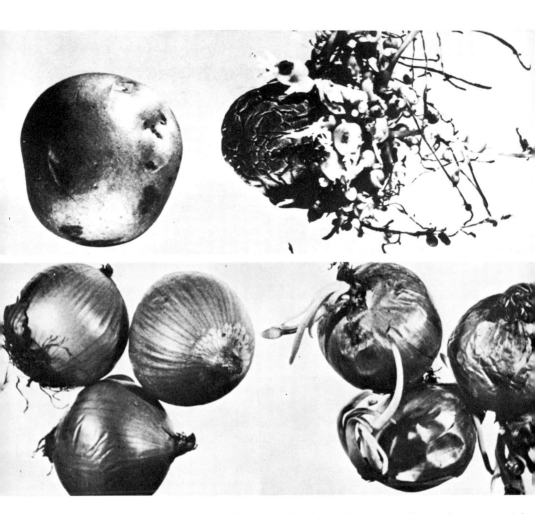

Food-preservation by radiation – after a year the treated samples are compared with untreated on the right

ducts. Its hard-wearing properties make a material particularly suitable for floor coverings and a new airport in Finland is applying it to floors throughout the premises.

An interesting application of this nuclear age material is in the construction of skis. Ten pairs, with a thin veneer of the material on the underside, have been made in Finland and some have been worn for over 500 kilometres of skiing without showing signs of wear. Many more applications will be developed for a variety of industries, such as sports goods, toys and tool manufacture. And there are possible uses in the textile industry for the production of crease-proof garments.

But the most striking success of radiation processing has been the sterilization of surgical appliances, like rubber gloves and catheters, which are difficult to sterilize by conventional methods. The appliances are sealed in air-tight containers before irradiation so that re-infection cannot occur. Pre-sterilized hypodermic syringes are produced so cheaply that they can be thrown away after one use. This has probably been the greatest boon to the medical practitioner in recent years.

The preservation of food by radiation has aroused a great deal of interest among food technologists, and the technique is already in use in several countries. Radiation treatment seems to be the only way of increasing the storage time of many perishable food products, like some fruits, which cannot be cooked or treated by other conventional methods. And irradiation will disinfect other foods, such as grains, without leaving debris. The shelf-life of fish and some meats can be increased sufficiently to permit them to be sold as nearly fresh products. Irradiation can also suppress sprouting in potatoes and onions, and delay the ripening of certain fruits, such as bananas and pears. Exhaustive tests have been made by human volunteers to establish the safety of irradiated foods. The foods are exposed to gamma-radiation from sealed radionuclide sources and, therefore, they are not at any time made radioactive.

Radionuclide techniques have been much used in the environmental and ocean sciences, particularly in the fields of hydrology, oceanography and meteorology. The study of water resources is an important application of great value in the developing countries, and radionuclides are used for investigating such matters as the amount of water flowing down a river, the direction and flow of underground streams, seepage from reservoirs and canals into the soil and the origins of springs and lakes. In countries like Israel, where water is at a premium, they have already been widely applied. Radionuclides are playing an increasingly active role in oceanography, a rapidly expanding science covering such diverse fields as ocean transport, mineral exploitation and sea-food production. One of the major uses so far has been the study of the movement of coastal silt, sand and pebbles. Radioactive tracers are extensively employed to investigate coastal and estuarine water movement, often in relation to sewage and chemical waste disposal. Tracers have also helped meteorologists further their understanding of atmospheric behaviour. They have been used, too, in measuring atmospheric pollution. Specific investiga-

Bottles of tritium-labelled water are dropped into Lake Chala, Kenya, to assess the potential of a spring-fed irrigation scheme in the proximity. If the spring-water shows an increased radiation count the link will be established

tions of sources of pollution have involved tracer studies of, for instance, the diffusion of fumes from chimneys.

Radionuclides have even led to an improvement in the quality of beer. By finding the life-time of radio-actively-labelled lactic acid bacteria it has been possible to reduce the fermentation process from five days to about one.

Even though usually unappreciated, radionuclides have for many years played an active, and often unique, role in our daily lives. But nuclear technology is young and fast moving, and much is expected from other applications which have yet to be fully developed. The most important of these are the use of nuclear energy for propulsion and the peaceful uses of nuclear ex-plosions.

In an important application of radio-nuclide techniques in oceanography a load of labelled silt is deposited in the sea and its subsequent movement followed by measuring the radio-activity of samples from the sea floor

MINI-REACTORS 5

At sea

The use of reactors for propulsion has always fascinated nuclear engineers and the first power reactor ever constructed was in fact built in a submarine hull at the National Reactor Testing Station in Idaho. It started operating in March 1953 and was the forerunner of the pressurized water reactor used in the first American experimental submarine, the *Nautilus*, launched in 1954. An extensive effort was also made in the United States to develop a high-temperature reactor for the propulsion of military aircraft; but in 1961 it was decided that 'the possibility of achieving a militarily useful aircraft in the foreseeable future is still very remote' and the project was dropped. It is probable that the development of missiles was the main reason for this decision.

The United States Navy now has a large fleet of nuclear vessels all powered by pressurized water reactors. The *U.S.S. Skipjack* is the lead ship in a fleet of nuclear attack submarines, the *U.S.S. George Washington* leads a fleet of ballistic missile submarines, the *U.S.S. Bainbridge* a fleet of nuclear destroyers, the *U.S.S. Longbeach* a fleet of nuclear cruisers and the 86,500-ton *U.S.S. Enterprise* is the first nuclear aircraft carrier and the largest warship in the world.

The tremendous attraction of nuclear power for naval propulsion is due primarily to the compactness of nuclear fuel and, for submarines, to the fact that oxygen is not required for engine operation. This

The U.S.S. Enterprise – the first nuclear powered aircraft carrier

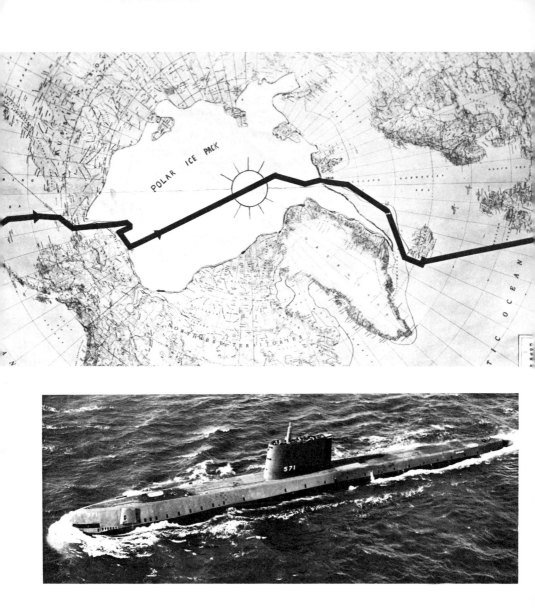

POLAR ICE PACK

The first nuclear powered vessel was the Nautilus which undertook this marathon journey without refuelling

means that nuclear ships in general have a relatively long range and high cruising speed and that nuclear submarines can remain submerged for very long periods. Conventional submarines, powered with diesel engines, can sustain a maximum surface speed of about 18 knots for only about 30 minutes. Underwater, they can travel for an hour at 8 knots and must then surface to recharge their batteries – submerged operation is only possible for about one-seventh of the time they are at sea. The characteristics of nuclear submarines are, by comparison, staggering. They can operate submerged almost indefinitely and can travel at speeds in excess of 20 knots for weeks at a time. Underwater, they can make more speed than on the surface and they have remarkable range. The *Nautilus* steamed for almost 100,000 miles on one fuel loading; and in 1960 the *Triton* made an uninterrupted submerged journey around the world, a distance of nearly 40,000 miles in 83 days.

Nuclear ships are also operated by the navies of Great Britain, France and the Soviet Union.

The possibility of using nuclear power to propel merchant ships was first seriously considered in 1954 by the United States. And the *N.S. Savannah* was built at the Camden, New Jersey shipyard of the New York

The 22,000 ton passenger-cargo NS (nuclear ship) Savannah which ushered in the use of nuclear power in passenger and cargo ships

Shipbuilding Corporation for the States Marine Lines to test the safety and reliability of a nuclear vessel carrying both passengers and a cargo. She was launched in July 1959 and satisfactorily completed sea trials in 1962. The ship displaces 23,000 tons, is 600 feet long, 78 feet in beam and is propelled by a pressurized water reactor, of the type used in the *Nautilus*, that delivers 22,000 shaft horse-power to sustain a cruising speed of 20 knots. The reactor core is designed to have a long lifetime and the uranium fuel contained in the reactor can produce enough energy for $3\frac{1}{2}$ years of normal operation without refuelling – equivalent to a cruising range of 300,000 miles. The power output of the reactor is varied by the automatic operation of boron control rods and the heat from the coolant is used to generate steam to drive the turbines on the main shaft of the ship. The *Savannah* can be taken from one-fifth engine power to full power in 30 seconds and it takes only 1 minute to go from full astern to full ahead. The cargo capacity is 10,000 tons, and there is accommodation for 60 passengers and a crew of 110.

In 1964, the nuclear-powered West German ore-carrier, the *Otto Hahn*, was launched. It has a displacement of about 26,000 tons, a speed of nearly 16 knots and is able to transport up to 15,000 tons of ore. She too is propelled by a pressurized water reactor which provides 11,000 shaft horse-power. The member states of the European Economic Community participated in the design and construction of the *Otto Hahn*, which is intended to be the forerunner of a European nuclear merchant fleet.

The Soviet Union launched a nuclear ice-breaker, the *Lenin*, as long ago as 1959 and plans to build a second one. The *Lenin* has three pressurized water reactors – two are used and one is held in reserve. The ship has a displacement of 16,000 tons and delivers 44,000 shaft horse-power to three screws. It has seen much service in the Arctic Sea and needs refuelling only at two-year intervals.

And Japan has recently launched a small nuclear vessel. Italy intends to do so soon.

'Lenin' – a Soviet nuclear powered ice-breaker on an ice reconnaissance journey.

Until recently, maritime experts doubted whether nuclear power would be economically justified to propel liners, cargo vessels or oil tankers. But now more and more shippers are becoming interested in the rapid transport of high-value goods to minimize the immobilization of the large sums of capital tied up in the cargo. Furthermore, shipowners are in a highly competitive business and are anxious to raise the profitability of their ships by reducing operating costs. In many ports, handling costs and dues are increasing continuously and it is important that the turn-round time of cargo ships is minimized. And it is also important that the time spent at sea is reduced.

Time in ports can be cut by the use of standard-sized containers to increase the efficiency of cargo loading and unloading. Therefore, nuclear propulsion is being considered for future high-speed container ships. The potential economic advantages are the elimination of fuel-oil tanks or coal bins, making more space available for cargo, and improved ship utilization resulting from higher cruising speeds and the very much less frequent refuelling. But, at present, these advantages are cancelled out by the high capital cost of nuclear ships. The balance should move in favour of nuclear propulsion for bulk-cargo carriers before it does for passenger liners or passenger-cargo combinations.

Recent studies have indicated that nuclear propulsion is likely to be economically competitive above about 40,000 shaft horse-power. For example, for long-distance ocean operation, a nuclear container-ship of 15,000 tons carrying 900 standard containers would probably be economically superior at speeds greater than 24 knots. Nuclear propulsion should also be competitive for supertankers of about 120,000 tons cruising at speeds of about 20 knots (40–50,000 shaft horse-power). And marine technologists are looking forward to the time when most cargo will be carried in very large nuclear submarines travelling under the North Pole.

The advantages of nuclear propulsion become increasingly attractive as the capacity of the ship increases. Whereas the costs of oil-fuelled propulsion increase directly with capacity, the costs of nuclear propulsion should increase by only one-half when the capacity is doubled. A fast container-fleet operating between the United Kingdom and Australia would, if oil-fuelled, require 7 vessels travelling at 27 knots (52,000 shaft-horse-power) to operate a weekly service, but 4 nuclear-powered ships of 120,000 shaft horse-power would provide the same service by travelling at 30 knots and sailing at 11-day intervals.

Compared with land-based reactors, marine reactors must satisfy stricter criteria concerning safety, mechanical strength, weight and compactness. And it is necessary that the life of the core should be longer, a difficult requirement for a small reactor. Elaborate safety precautions have to be taken to protect the passengers and crew and to ensure that the inhabitants of ports of call are not endangered. All of the waste radioactive liquids from the reactor must be stored in tanks and none discharged at sea. It is not enough to guard against the possibility of a major explosion; strict measures must be taken against the danger of radioactive contamination caused by collisions or by the sinking of the ship in shallow waters.

In space

The nuclear rocket will play an important role in future plans for space travel and space operations. The main attraction lies in its potential for operation with a specific impulse, the pounds of thrust per pound of propellant ejected per second, two or three times greater than that of a chemical rocket.

Nuclear rockets appear to offer the greatest advantage in future manned missions to the planets. A nuclear-propelled spacecraft built for a journey to Mars would weigh only one-tenth as much as an equivalent chemically-propelled vehicle. The round trip, which might be carried out in the late 1980s, wou d take about 18 months. And it is probable that only by the exploitation of nuclear fuels will it be possible to obtain sufficient independence from the earth to visit other planets.

In a nuclear rocket, liquid hydrogen would be vaporized and the gas heated to a high temperature in a reactor. The hot gas would then be ejected by expanding it through a rear nozzle, thereby developing thrust. Extremely high reactor temperatures and, therefore, very large power outputs would be required for efficient rockets. And the reactor must be able to start and stop quickly and precisely. The National Aeronautics and Space Administration and the USAEC are jointly sponsoring a programme to develop nuclear rocket engines for space missions. In the Nerva project (Nuclear Engine for Rocket Vehicle Application) an engine, 22 feet tall and providing a thrust of about 50,000 pounds at a high specific impulse, is based on a graphite reactor.

The AEC is also developing a reactor for electric propulsion. This is, so far, an unproven concept for space vehicle propulsion in which thrust will be produced by ejecting a high-speed beam of electrically charged particles instead of a gas. The particles could be either ions or electrons. Theoretically, very high specific impulses, of the order of ten thousand pounds of thrust per pound of matter ejected per second, could be obtained by this method. The snag is that the matter would be ejected only slowly, at a rate of a few thousandths of a pound per second and so electric propulsion is potentially a low-thrust system. But it could be very useful for manoeuvring or propelling spacecraft that have been taken out of the earth's gravitational field by other propellants. Nuclear reactors offer the best means of generating the electrical power needed to ionize and accelerate the charged particles.

Auxiliary power supplies

An important prerequisite for deep-space exploration is a compact auxiliary power system, to operate instruments and other devices, which is also long-lived, reliable, mechanically strong and lightweight. At present, spacecraft power supplies depend on the use of

Designed for use in space this nuclear reactor has no moving parts. Without human adjustment it generated electricity for a year in the vacuum chamber (below) which simulates space conditions

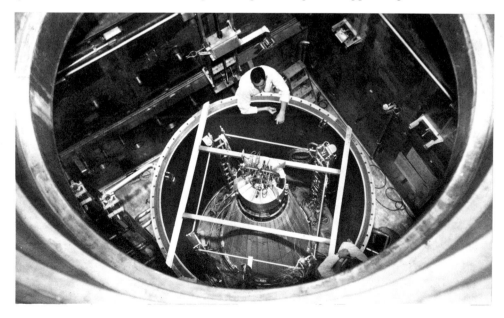

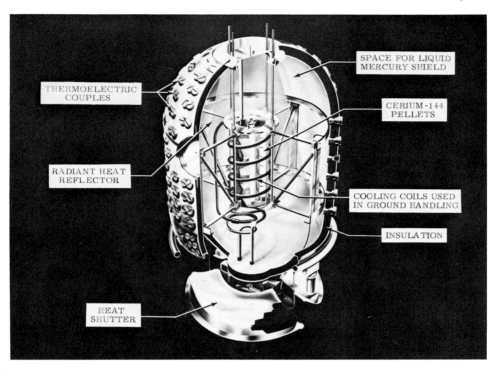

THERMOELECTRIC
COUPLES

SPACE FOR LIQUID
MERCURY SHIELD

CERIUM-144
PELLETS

RADIANT HEAT
REFLECTOR

COOLING COILS USED
IN GROUND HANDLING

INSULATION

HEAT
SHUTTER

The SNAP generator showing how it works and placed at the astronaut's feet during the Apollo-12 mission

solar cells for generating electricity for direct use and for recharging chemical batteries. But, as power requirements increase, the design of larger solar cells and batteries becomes much more difficult. And nuclear power supplies would be essential for missions travelling great distances from the sun – the large size of the solar cells required makes them impracticable. Nuclear generators do not require sunlight. Even on lunar missions they offer considerable advantages. With the moon in darkness for 14 days in every 28, solar batteries can operate for only 50 per cent of the time. But nuclear generators operate all of the time.

The first application of nuclear energy in space was, in fact, the use of a nuclear battery, a device that generated small amounts of electricity by the conversion of heat given off by a radionuclide as it decayed. In 1961, a nuclear battery, weighing 2 kilograms and generating 2.7 watts of electricity, was taken into orbit around the earth on a US-navy satellite. Its purpose was to provide the power for two of the satellite's navigational transmitters which provide a means for ships and aircraft to determine their positions at any point in the world.

A radionuclide thermoelectric generator was taken to the moon by Apollo 12. The nuclear assembly was carried on the outside of the lunar module on the journey, so that the heat generated was dissipated in space. The unit was designed to produce over 60 watts of power to operate instruments and systems which are continuously sending information back to earth. It is cylindrical in shape, about 18 inches high and 16 inches across, made of lightweight beryllium and weighing about 28 pounds, excluding the fuel. The fuel capsule is 16.5 inches long and 2.5 inches in diameter, and weighs about 15.5 pounds, of which 8.4 pounds is fuel. Plutonium-238 is the radionuclide, and a thermopile assembly converts the heat produced by radioactive decay directly into electrical energy. There are no moving parts.

The properties of Pu-238 make it an ideal isotope for space generators. It still provides half of its original heat after 90 years, and since it emits mainly alpha particles it requires no extra shielding – the generator casing essentially stops all the escaping radiation.

Nuclear batteries also have numerous potential applications on earth for such special purposes as the provision of power in weather stations and lighthouses.

An intriguing use of small nuclear power supplies is in heart pacemakers. Pacemakers have been in use for about ten years to maintain the heart beats of patients suffering from a disease known as heart-block, caused by the failure of a bundle of nerves (the bundle of His) in the heart. A pacemaker is implanted into the patient to provide the minute rhythmic electrical pulses normally passed through the nerve bundle. Ordinarily, the devices are powered by small chemical batteries but the snag is that these need replacing every two years or so which means repeated surgery for the patient. Nuclear-powered pacemakers are being developed which will have a lifetime of more than ten years so that their use will avoid this repeated surgery. The nuclear battery utilizes the heat from the decay of Pu-238 to generate electricity in a minute thermoelectric generator. The complete battery is 2 inches long and about half an inch across, and weighs about an ounce. The battery is sealed so that radioactive material cannot get out and body fluids cannot get in. Initial trials in which nuclear pacemakers have been implanted into dogs, have proved successful and they have already been implanted into a small number of humans. Their use should ensure that many heart-block cases will live normal lives without the inconvenience of frequent hospitalization.

Large auxiliary power sources, based on extremely compact reactors, are being developed. Their power outputs will vary from 500 watts to hundreds of kilowatts. One of these reactors is liquid-metal cooled and drives either a mercury vapour turbogenerator, giving out between 3 and 150 kilowatts, or a thermoelectric generator supplying between 0.5 and 50 kilowatts. A reactor system of this type, producing 500 watts, has been launched and tested in orbit. Its operational lifetime is about one year and it weighs about 950 pounds.

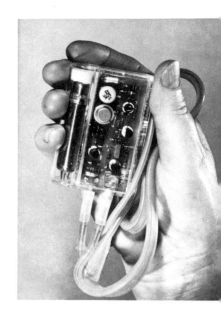

One handful of power that, implanted into the body, will support the life of a cardiac patient for ten years without repeated surgery. This cardiac pacemaker is powered by an alpha-emitting radionuclide and initiates the heartbeat with a regular pulse

Right: deep in the Antarctic a small reactor provides heat, light and steam for water at a research camp, eliminating the problem of hauling thousands of barrels of fuel across the snow cap

Below: a RIPPLE radionuclide power source developed in the United Kingdom for a number of applications where economy of space, long life, reliability and minimum maintainance are essential

A larger system, producing 35 kilowatts, has also been tested in space. This one is designed to last for 10,000 hours and weighs about 3,000 pounds.

Compact reactors

In the foreseeable future it will not be possible to make nuclear reactors small enough to drive directly lorries, cars or other small vehicles. But it will be possible to use reactors to produce chemical fuels for vehicles, from substances like air and water which are readily available everywhere. For example, hydrogen from water can be combined with nitrogen from the air to produce ammonia which can then be 'burned' in internal combustion engines and turbines. In the 'energy depot' concept a reactor will be brought into an area of fuel shortage together with a fuel manufacturing plant to become a combined refinery and fuel station. Vehicles in the vicinity will then drive in and fill up with the manufactured fuel.

Very small power reactors are already in operation for other purposes. A 'portable' reactor has been used since 1962 by the United States Air Force to power and heat a radar base on a remote mountaintop in Wyoming. It was transported there by cargo transport planes in 27 packages, each measuring 8 feet by 8 feet by 30 feet and weighing 30,000 pounds. The pressurized water reactor has an output of 1 MWe. Two other similar portable reactors are in operation, one in Antarctica and one in Alaska. They are particularly useful in these remote and frigid areas to supply power and heat to military and scientific bases, hundreds of miles from sources of conventional fuels. A fourth plant is mobile, mounted on a converted American Liberty ship called the *Sturgis*. This reactor can operate at full power, producing 10 MWe, for one year without re-fuelling – an equivalent diesel-powered plant would require 160,000 barrels of oil per year. The *Sturgis* can be used to supply power in disaster areas near the coast

Below: part of a small nuclear power reactor is loaded into a transport air-craft for removal to a remote area. The entire reactor can be moved in a few airlifts

Top: The 'Sturgis', installed with a nuclear power reactor can be moved to any coastal region to provide electric power. Bottom: a Soviet motorized nuclear power plant

where floods, hurricanes or earthquakes have knocked out normal power supplies. The plant would be anchored offshore and power taken to the land through submarine cables or overhead wires. A design for a desalination plant, producing 500,000 gallons of fresh water per day, to be installed on the *Sturgis,* has been developed so that it can supply water as well as power to the disaster areas.

NUCLEAR EXPLOSIONS IN PEACE   6

At the first underwater nuclear test at Bikini Atoll in 1946 a fleet of obsolete and surplus naval vessels was sunk in a dramatic demonstration that an extremely powerful shock wave could be propagated in condensed matter by a nuclear explosion. And the Eniwetok test of a thermonuclear weapon showed that a large nuclear explosion could vaporize a Pacific Atoll, leaving a cavity where an island had previously existed. These incidents gave birth to the idea that nuclear explosions could also be used for a variety of peaceful purposes and, over the past 15 years, the concept has been given serious study, particularly in the United States. Progress in the field has been stimulated by advances in the theory of explosion effects and by the development of thermonuclear explosives having only a small fission component. These 'cleaner' explosives have somewhat lessened the potential radioactive contamination from nuclear explosions. But this decrease is only relative, since all nuclear explosives produce radioactive isotopes.

Plowshare

A formal programme for the peaceful applications of nuclear explosions was begun in the United States in 1957 and given the euphemistic title 'Plowshare', derived from the biblical phrase 'They shall beat their swords into ploughshares; neither shall they learn war any more.' In the first four years, a series of experi-

ments was performed with chemical (non–nuclear) explosives to study the effects of large underground explosions. The first nuclear Plowshare test, 'Project Gnome', took place in 1961 and has been followed by many others. The experience accumulated from these has led to such specific proposals for applications on a commercial scale that a consortium of American and European countries has been formed to promote the peaceful uses of nuclear explosives in projects like the excavation of canals and the building of tunnels, the recovery of natural resources (oil, gas and minerals), and the building of harbours and dams. And in the Soviet Union, because the amount of construction work to be done is so great, and the mineral and petro-chemical deposits to be tapped are so large, the idea of

Left: A huge column of water surges into the air over Bikini Atoll to form the familiar mushroom over a target of 83 obsolete warships in the first underwater detonation of a nuclear explosive

Below: a hook-shaped tunnel leading deep into a mountain in which a nuclear device was detonated as part of the Plowshare project to assess the potential of nuclear explosives in civil engineering works

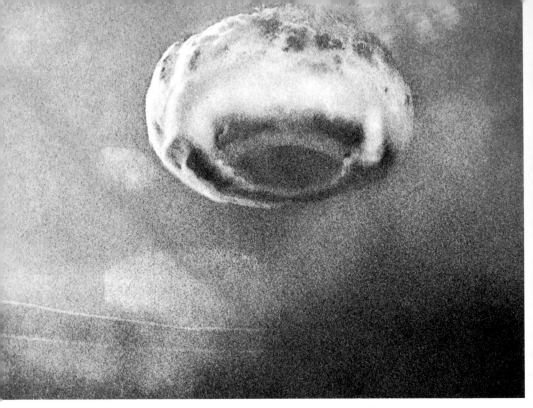

Above: a fireball produced by a thermonuclear explosion which, in the atmosphere, disperses causing relatively little radioactivity. Detonated underground, as in the Gnome experiment, a huge cavity (opposite) is formed with glassy radioactive walls of melted rock which subsequently collapses into the base

using nuclear explosives for these purposes is known to be particularly attractive.

Almost all the energy released in a nuclear explosion is given off within a few millionths of a second. A fireball is formed in which temperatures of millions of degrees and pressures of millions of atmospheres exist. If the explosion occurs underground the surrounding rock is vaporized and a cavity, with molten rock boundaries, several metres across is formed. A few thousandths of a second after the explosion the shock wave strikes the wall of the cavity, causing it to expand further, and travels on until its energy is dissipated, crushing and fracturing the surrounding rock. Within a few seconds, the molten lining of the cavity flows to the bottom and solidifies to a bowl-shaped mass which contains most of the radioactivity of the nuclear explosion. If the explosion is deep underground, the crumbled rock above the cavity is insufficiently solid to form a roof and some of it falls in so that the cavity

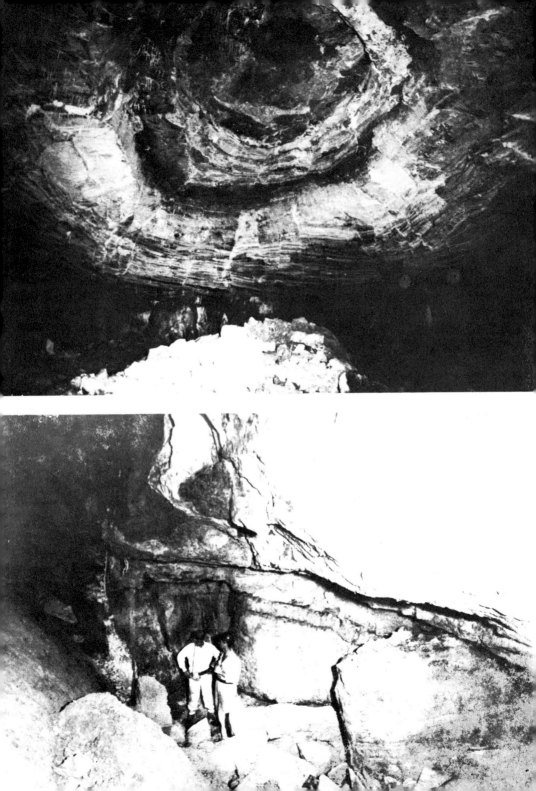

becomes filled with loose rock. And rock will continue to fall until a point is reached at which the less shattered rock is strong enough to form a bridge to support the rock above it. This results in a cylinder filled with crumbled rock, called a 'chimney'. No effects can be seen at ground level if the explosion is very deep but, if it is less deep, the chimney formation will cause the earth to sag and, if it is less deep still, the chimney could extend to the surface, causing ground cracks.

The potential advantages of nuclear explosives, compared with chemical ones, are cost and size. A 25-kiloton nuclear device – equivalent in explosive power to 25,000 tons of TNT – could be emplaced in a cylindrical bore-hole less than a metre in diameter whereas this quantity of TNT would require a spherical cavity 30 metres across. Even if the TNT could be assembled in one place, the cost of emplacement would be enormous. Yet even a one-megaton nuclear charge, equivalent to 1,000,000 tons of TNT, could be accommodated in a cylindrical hole not much more than a metre in diameter. Moreover, nuclear explosives are relatively cheap. The USAEC envisages a charge of $350,000 for a nuclear device equivalent to 10,000 tons of TNT and $600,000 for 2,000,000 tons. The costs of the equivalent TNT would be $4 million and $800 million respectively. Large explosions using chemical explosives are simply not economically feasible.

In December 1967, at the San Juan Basin in New Mexico, a 26-kiloton nuclear explosive was detonated at a depth of 4,240 feet in a rock formation. The rock was gas-bearing shale and the project, called 'Gasbuggy', was to study the feasibility of increasing the output of a low-producing natural gas field. Normally, gas is obtained by drilling a well into the rock formation and then pumping out the gas forced into the well by natural underground pressure. But the region around the well that can be tapped in this manner is often limited because the natural gas is held in comparatively non-porous rock that prevents the flow of all but small volumes of it into the well. Conventional methods of increasing the flow of gas involved either forcing fluid under high pressure into the well or using underground chemical explosions to fracture the rock and form pathways for the gas. But the result is usually short-lived and not worth the extra cost.

The USAEC predicted that the Gasbuggy explosion should produce a cavity 160 feet in diameter and produce such extensive cracks in the rock that as much as 75 per cent of the gas from a wide zone around the site of the explosion should flow into the well over a 20-year period. The chimney formed was 330 feet high, with a volume of at least 2 million cubic feet, and pathways for the gas radiated out for distances of hundreds of feet.

Samples of the gas freed by the nuclear explosion have been tapped off from the top of the chimney and tested for radioactive contamination; and it seems that the radioactivity will eventually be low enough for the gas to be used commercially without much purification. During a series of tests lasting 90 days, a total of 109 million cubic feet of gas was produced. By contrast, about 400 feet away there is a well in the same rock formation created by a conventional method, which has produced 85 million cubic feet in 9 years.

In September 1969, a second nuclear explosion, 'Project Rulison', with a 40-kiloton yield, was detonated 8,400 feet below ground, at a site near Rifle, Colorado, in a similar experiment. If Gasbuggy and Rulison are successful, they will be the forerunners of many subterranean nuclear explosions to tap natural gas that is otherwise irrecoverably lost beneath the earth's surface. For example, the U.S. Bureau of Mines has estimated that nuclear explosives, if proved practicable, would more than double the present natural gas reserves in the Rocky Mountain area alone.

Petrochemical and mineral recovery

It has been estimated that over 90 per cent of the world's petroleum is to be found in oil shales. If the reservoir of oil locked in shales can be tapped, our reserves of fossil fuels would be greatly increased. The conventional technique is to dig down and heat the under-

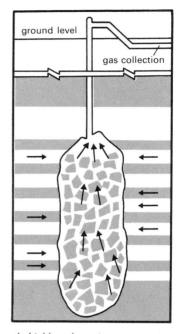

A highly schematic representation of the chimney caused by the Gasbuggy explosion designed to release natural gas. The shaded areas represent layers of shale

ground shales and tar sands so that the oil can be drawn out and collected. But it has been suggested that a better method would be to detonate a nuclear explosion to break up large quantities of shale or melt huge amounts of tar sands. It would then be easier to retort the oil out of the permeable oil-bearing material. This concept is still speculative and no test has so far been performed but the financial pay-off for successful development would be very large indeed. For example, it has been estimated that 400 million barrels of oil are potentially recoverable from a formation in Colorado and it has been proposed that a 50-kiloton nuclear explosion should be set off at a depth of 3,350 feet in the Piceance Creek Basin in an attempt to do so.

Another possible use of nuclear explosives is in the mining of metallic ores such as copper. After copper ore has been mined, the metal is removed by dissolving the ore in acid and separating the copper from the solution by a chemical process. If a nuclear explosive was detonated to break up the ore over a large area the acid could be sent underground and the solution containing the copper pumped out. This would greatly simplify the procedure since the only operation performed on the surface would be the removal of the copper from the solution leaving the mine. Provided that the underground ore is made sufficiently permeable this application is an attractive alternative to deep mining for such metals as copper and uranium. Many of the radioactive contaminants produced by the explosion would be removed by the chemical processing of the ore solution. 'Project Sloop' is a proposal to test this concept by detonating a 26-kiloton nuclear device 1,200 feet underground near Safford, Arizona.

Nuclear civil engineering

The most familiar peaceful use of nuclear explosions is for earth moving in major civil engineering works, including the building of harbours and canals, and the division of waterways. For these applications, the explosives would be detonated at much shallower depths than for mineral and petrochemical extraction.

Once again, the tremendous energy of the explosion forms a cavity underground but the portion of earth above it is lifted into the air causing the sides of the cavity to cave in somewhat and leaving a conical crater surrounded by a lip of crushed loose rock. The molten and resolidified rock zone, containing most of the radioactivity, still exists but is buried under the crater floor. If a series of suitably placed nuclear devices are detonated simultaneously, the craters would overlap to form a ditch. Thus, in 'Project Buggy', a nuclear ditch 20 metres deep, 75 metres wide and 260 metres long was produced in hard rock by detonating five 5-kiloton nuclear charges at a depth of 40 metres and at appropriate spacings. The possibility of using the nuclear ditching technique as a cheap way of building a new sea-level canal across the central American isthmus is being considered. The Panama Canal as it now exists is not at sea-level but is a lock canal. And very large vessels, such as aircraft carriers and oil tankers, cannot navigate it. It would cost about $2 billion to enlarge the canal and take it down to sea-level by conventional methods; but the cost would be much less if nuclear explosives were used. In practice, the use of nuclear explosives along the present route is ruled out because the main population centres in Panama are located close to the canal – the radioactive contamination and earth shocks would be far too great. But an alternative route, in another stretch of Panama, mostly jungle and sparsely populated, has been considered. To blast this 100-mile long ditch would require 300 nuclear explosions, equivalent to a total of 200,000,000 tons of TNT, at a cost of about $1 billion. The result would be a much wider and deeper canal than could be made by conventional methods. But, to be sure that no person was endangered by radioactivity, it would be necessary to evacuate 25,000 inhabitants from an area of 10,000 square miles.

The detonation of about 250 nuclear explosives, at appropriate spacings and at depths of between 150 and 185 metres, is part of a massive Soviet project to build a 112-kilometre canal across the Pechora-Kolva water-

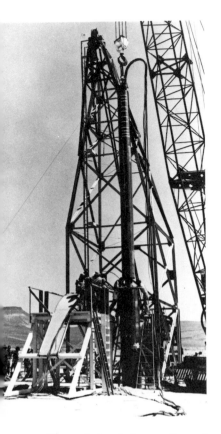

shed. The explosions will excavate a nuclear ditch, 65 kilometres long, through the rocky section of the canal route – the remainder will be constructed by conventional hydraulic engineering. The sparseness of the population (less than one inhabitant per square kilometre) around the site of the proposed explosions makes the project feasible. By nuclear digging the costs of the canal should be slashed by a factor of about 3 compared with conventional methods. The ultimate aim is the diversion of some northern rivers into the Volga to stabilize the level of the Caspian Sea. Since 1935, this has fallen 2.5 metres due to abnormal weather conditions and increased water usage, with harmful effects on fishing and much inconvenience to sea transport. And it appears that the level could drop a further 1.7 metres by the year 2000. The river diversion will also satisfy the growing demands for water in the central and southern regions of the European part of the Soviet Union.

The explosive device, eruption and crater of Project Sedan. That the comparatively small explosive, seen above with associated equipment, should provide the energy to move 8 million cubic yards of earth (right) in a few seconds gives an indication of the enormous power potential of nuclear energy

In the United States in 1962, a 100-kiloton nuclear explosion was detonated, at a depth of 190 metres, in Project Sedan. The explosion was calculated to give a crater of maximum size, which turned out to be 100 metres deep and 350 metres in diameter. About eight million cubic yards of earth and rock were displaced and most of the radioactivity was trapped in the cavity, 90 metres below the crater floor. The majority of the remainder was on the surface of the broken rocks on the crater floor but some escaped into the atmosphere. Explosions of this type could be used to construct harbours in places where the rock floor of the sea is not deep enough for large ships to dock and have already been considered for certain sites in Australia and Alaska.

At the lower end of the scale of possible civil engineering applications is straightforward quarrying – in which a considerable amount of rock would be fractured to make its removal easier in, for example, railway and road construction in mountainous regions.

For dam construction in isolated areas, a cratering explosion could be used to shift large amounts of rock to the point where a stream is to be blocked. And the nuclear explosions would also produce broken rock for the production of concrete aggregate.

Nuclear explosions could also be used to create underground storage areas. It has been estimated that a 100-kiloton explosion would form a reservoir capacity exceeding 350,000 cubic metres. If detonation took place at the proper depth in the right type of rock the reservoir could be used to store water, oil, natural gas or other fluids.

In the arid sun-baked regions of the Central Asian republics the acute water shortage is restricting the settlement of the virgin lands. And, on the basis of experience gained in experimental cratering shots, Soviet planners are looking to nuclear explosives for a rapid and appropriate solution to their irrigation problems. Underground nuclear blasts will create a reservoir complex, of 30 million cubic metres capacity, to collect enough of the spring run-off waters to provide eventually nearly 80 per cent of the water requirements of the region. In the first stage, two 150-kiloton nuclear explosions at a depth of 185 metres will create barrier dams. Construction gangs will move in two months after the explosions and complete the job in five months. It seems that the radioactive content of the reservoir water will be sufficiently low to allow its use, even for drinking.

Scientific applications

Peaceful nuclear explosions have certain scientific applications, particularly in nuclear physics. In some of the very large nuclear-weapon test explosions in the Pacific it was discovered that new artificial elements, heavier than uranium, were produced by the neutron bombardment of uranium. These transuranic elements were found in the debris of the explosions. Some nuclear physicists, interested in producing and studying them, would like to use underground nuclear explosions because they are the most powerful means of obtaining a large number of neutrons in a short time.

Nuclear explosives can also supply a great deal of geophysical information, and much of our knowledge of seismology and of the nature of the earth's crust has been obtained from measuring seismic effects after underground nuclear explosions.

Although we now know a great deal about nuclear explosions there are still many problems to be evaluated in any specific application. We only have imprecise understanding of the effects of the explosions and how they vary with amounts of explosive, depth of detonation and types of rock. Qualitative assessments can be made of the costs and benefits of applications like canal and harbour construction, but many more years of research and development will be needed before they are developed to a stage where they can be widely used for peaceful purposes. And parallel research investigations in the related fields of seismology, ecology and geology will be necessary to provide a complete knowledge of the side effects of the explosions. The possible spread of radioactive contamination is, of course, a major consideration. For example, the consequences of the release, over a long period of time, of radioactivity trapped in harbour and canal digging projects require detailed examination.

Radioactive contamination

At least four of the Plowshare underground test explosions have accidentally released radioactivity into the atmosphere. In April 1965, the 'Panquin' test, in which a 4-kiloton device was exploded at a depth of 280 feet, burst through the surface of the earth, releasing much radioactivity into the atmosphere. In the 'Cabriolet' test of January 1968, radioactive debris was hurled 1,900 feet into the air. And the 'Schooner' test of December 1968 released into the atmosphere the highest levels of radioactivity recorded in the western areas of the United States since the treaty banning tests in the atmosphere came into force in 1963. The treaty allows underground tests which release radioactivity into the atmosphere, provided it does not cross the frontiers of the state in which the test takes place. But the radioactivity from the Schooner test was detected by sampling stations in Canada, and the Mexicans reported radioactivity in their atmosphere after an underground test in January 1969. So it appears that the Plowshare tests have contravened the test-ban treaty on at least two occasions.

It has been suggested that underground nuclear explosions might trigger earthquakes. Even though it seems clear that they do trigger a large number of small earthquakes within a few miles of the test, most of these are very much smaller than the test itself. All planned tests are considerably smaller than the largest earthquakes that occur naturally. Whether or not earthquakes are triggered at large distances is not known with certainty, but there is some evidence that the larger tests can.

A threat or a promise?

There are many authorities who, while readily admitting the increasing importance of the application of nuclear energy to electricity generation and to medicine, agriculture and industry, firmly deny the value of peaceful nuclear explosives. They argue that Plowshare has been much over-advertised in the United States by certain politicians influenced by a number of enthusiasts and by commercial interests. Nuclear explosives, they say, can do nothing of importance that cannot be done by non-nuclear methods at greater labour and expense but without the problem of radioactive contamination. Objections are also raised because a nuclear explosive device designed for peaceful purposes could also be used as a nuclear weapon.

This issue was discussed at length by the Eighteen-Nation Disarmament Committee during the negotiations for the non-proliferation treaty. During the debate, certain non-nuclear-weapon states argued forcibly that they should be allowed to manufacture nuclear explosives for peaceful purposes. Senor Correa da Costa, the Brazilian representative, made a particu-

One of the possible effects of underground tests is to cause earthquakes. Here a seismologist studies subterranean conditions to ascertain the effects of a test

larly passionate speech in which he stated that his country had no intention of acquiring nuclear weapons but would not waive the right to conduct research and manufacture 'nuclear explosives that will enable us to perform great engineering works, such as hydrographic basins, the digging of canals or ports – in a word, the reshaping of geography if necessary, to ensure the economic development and the welfare of the Brazilian people'. He referred to an 'ever-increasing variety of applications' and the 'boundless prospects' of nuclear explosives 'that may prove essential to speed up the progress of our peoples'. There is no doubt that the Brazilian somewhat overstated the case for peaceful nuclear explosives. At least five years of research and development will be needed before they can be used for even the most straightforward projects and only a very few of these will be completed by 1980. But there is a high probability that, by the end of this century, nuclear engineering will have become a standard, though not a routine technique.

THE NUCLEAR FUTURE

The first step into the nuclear age was the establishment of nuclear power as a safe, reliable, energy source, competitive with fossil fuels – an achievement which required a quarter of a century of research and development. The second step will be taken when reliable advanced breeder reactors are developed to the point of producing very cheap electricity in large quantities. And it will be very surprising if this takes much less than another quarter of a century. The final step will be the widespread application of this cheap energy to many industrial and chemical processes. The whole process will have probably taken one century from the time, in 1942, when Enrico Fermi established the first chain reaction at Chicago – a rather longer period than was anticipated in the 1953 'Atoms for Peace' programme.

By the year 2000, the total energy requirements of the world are likely to be more than four times greater than they are at present – about 40 per cent of this energy will be produced by oil, 20 per cent by nuclear power, 20 per cent by coal, 18 per cent by natural gas and 2 per cent by hydro-power. Half of the increase will probably be due to the increase in population and the other half to the increase in per capita consumption. And the total world demand for electricity will probably be about 10 million MWe, and the annual rate of installation of new generating capacity about 500,000 MWe. It is likely that nuclear power will produce

The trail of a Minuteman II – these form the backbone of the US land-based strategic-missile force

nearly one-half of this electricity and that the rate of installation of new nuclear capacity will be about 400,000 MWe per year. These staggering figures imply that the equivalent of one nuclear power plant with a capacity of about 1,000 MWe will be brought into operation, at some location in the world, every day. These stations will almost certainly use very large advanced breeder reactors of several thousand MWe capacity and will produce electricity at a cost of 2 or 3 mills per kWh or less (at present prices).

But the energy gap between the developed and the developing countries is unlikely to be narrowed during this century. And, on present trends and indications, it appears doubtful whether the developing countries will even reach the existing energy consumption levels in the industrialized countries during this time.

From the experience of the developed countries it appears that nuclear reactors are only economical when used in sizes of at least 500 MWe. In countries like Chile, Israel and Greece the total capacities of the electric grid systems are less than about 2,000 MWe, so they cannot install nuclear stations large enough to compete with the conventional ones. If nuclear power is to be used to accelerate development, and in many cases it may well be the only thing that can, some method must be found to absorb the power from units of large capacity. One way would be to create the demand for electricity and for the nuclear station as a planned entity, a requirement that has given rise to the fascinating concept of the nuclear energy-centre.

Nuclear energy-centres will possess many of the characteristics of the vast complexes already established to exploit certain natural resources – for example; the SASOL complex in South Africa based on cheap coal; the aluminium complex in Kitimat, Canada based on water power; and the Tennessee Valley Authority development also based on water – but they will have one exceptional advantage: being independent of natural fuel resources, they can be established at locations where they will make the greatest economic and social impact.

When abundant supplies of low-cost energy are available the way that many industrial tasks are accomplished can be changed. There are a number of processes which can then be profitably undertaken by electrochemical methods, including the production of aluminium, magnesium, steel and pig iron. Cheap energy, so to speak, itself becomes a basic raw material, replacing other raw materials. Using a large nuclear plant, it would thus be possible to base many chemical and industrial processes on the direct use of low-cost electricity. These could then be located near to the sources of the basic raw materials or close to the markets for the products rather than in an area of cheap fossil fuels. This would greatly benefit many developing countries, because they could then decrease exports of ores and produce refined metals and chemicals actually at the mines.

The production of fertilizers – ammonium nitrate, urea and phosphates – is also an energy-intensive operation which could be done cheaply if large supplies of low-cost energy were provided. In several areas of the world, crop yields, even though already substantial, could be significantly increased solely by the application of more fertilizer. For this purpose, a large nuclear-powered fertilizer factory would be a profitable investment.

The energy-centre would also efficiently produce desalted water for irrigation. The major shortage in most of the hungry regions of the earth is not land but fertile, watered land. In fact, the amount of warm, potentially fertile and accessible, but arid, land is several times larger than the amount now used for producing food. Much of the land requiring irrigation is close to the sea. An agricultural system based on a controlled, pure and dependable source of water in a good climate, where all-year-round farming, producing three or four crops per year, is possible, may be economically feasible even in the most arid desert regions. The intensive farming would enable the maximum use to be made of capital invested in land and equipment and would eliminate the seasonal limitations of food production. An energy-centre producing electricity, water and fertilizer, often called an agro-industrial complex, is a particularly attractive proposition for many developing regions.

Suitable sites for such a complex exist in Chile, Peru, Mexico, some countries on the Mediterranean Sea, Pakistan and North Australia. It could be based on, say, a 3,000-MWe reactor producing about 1,000 million gallons of fresh water per day, with crops grown on a farm of about 300,000 acres – the actual choice of crops depending on local conditions. Several crops can produce a diet of 2,500 calories per day – adequate for one person – for an average water usage of 200 gallons per day; and so the complex could support a population of several million. High-value crops, such as citrus and cotton, could also be grown for sale in nearby markets. Considerable advances have been made in developing both new crop strains for intensive farming and evaporators for large-scale de-salination. And the combined technologies of agriculture, desalination and breeder reactors will no doubt make nuclear agro-industrial complexes viable. But numerous problems must be overcome before they can become commonplace. Not only are there the social problems associated with the establishment of large new communities in desert areas, but farmers

A nuclear energy centre could provide power for desalination, electricity generation and fertilizer production for large agricultural complexes

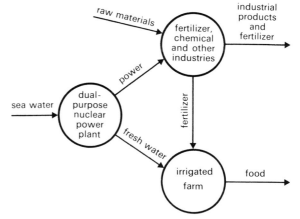

will have to be trained in new techniques; the water and fertilizer needs of crops under desert conditions will have to be determined; and the new enterprises will have to be financed and administered. Therefore, their introduction on a significant scale is not likely to be rapid.

Some developing countries are already actively considering the use of nuclear desalination. The development of nuclear power in Israel, for example, is closely linked to it. The installed electrical capacity in Israel is only about 1,300 MWe and the largest single unit that can be accommodated in the system is about 300 MWe. Nevertheless, the installation of a 1,300 MWe reactor is being considered to provide 300 MWe of electrical power and 1,000 MWe to produce about 100 million gallons of fresh water per day.

But the first reactor to be used for desalination is in the Soviet Union, at Shevchenko. This is a breeder – the largest in operation – with a capacity of 150 MWe and sited about 3 kilometres from the sea. Fifty million gallons of fresh water are produced per day. The largest conventional desalting plant in operation, in Kuwait, produces only about one-half of this output.

Nuclear power will also be used for water pumping. For example, the best source of water on the Ganges plain in India comes from underground wells; and year-round crops are possible, provided that cheap power is available for individual farm pumps.

Because each developing country has its own special conditions and problems it is not possible to predict a general pattern for the use of nuclear power for development. But it is certain that, when breeder reactors arrive on the scene they will be increasingly used for: producing power for energy-intensive industries at the site of the raw materials or close to the market; producing fresh water from sea water, inland brackish water or salty lakes; lifting and transporting water stored from seasonal rains; and producing huge quantities of fertilizer. We are moving into an era in which even the most arid lands will be made productive by combining with the abundantly available sunlight all the energy and water necessary for the efficient operation of the photosynthetic process – the basis of all food production and, indeed, of all life.

In due course, nuclear energy-centres will bring tremendous benefits with very wide-ranging implications. But the financing of the nuclear power programmes of developing countries will become a critical problem. So far, no nuclear plant has been financed from international sources, but four projects have been financed bilaterally under favourable conditions. Two power reactors at Tarapura, India, were financed with a United States aid loan, with a $\frac{3}{4}$ per cent interest rate and a 40-year repayment period. And the foreign exchange component of the Kanupp nuclear station in Pakistan was supplied by Canada, partly by a grant and partly by a 6 per cent loan. The Attucha reactor in Argentina will be financed by West Germany with a 6 per cent loan and a 25-year repayment period. But it is not at all certain that similar bilateral loans will be forthcoming in the future.

In the industrialized countries, dual-purpose nuclear plants producing water and cheap power will be used in metropolitan areas – first in the United States. Several cities, like New York and Washington, are capable of absorbing the 2,000 MWe and the 1,000 million gallons of water per day that will be produced by future plants. And nuclear energy will be used to produce heat to purify biologically contaminated but chemically acceptable water. The water supplies of many cities in the United States, Europe and Asia are of this type.

A future problem in many industrialized countries will be finding sites for new nuclear plants, with convenient, adequate and acceptable supplies of cooling water – already a difficulty in the United States and the United Kingdom. And, as the plants get larger, it will become more difficult. Over the next 20 years, about 200 new sites for large nuclear plants will be needed in the United States alone. The location of power centres will soon have to be decided by the ease of disposing of the surplus heat. And there are legal and other

Below: the distribution of world land conditions showing the vast amount of fertile land that could be put to use if sufficient power and water were available

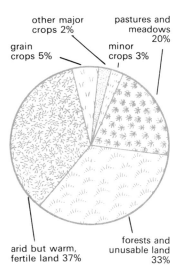

other major crops 2%

grain crops 5%

pastures and meadows 20%

minor crops 3%

arid but warm, fertile land 37%

forests and unusable land 33%

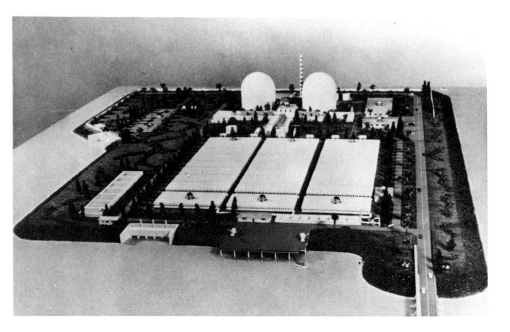

restrictions which will further reduce the number of available sites.

Many of the areas requiring large supplies of energy will be on, or near, the coast. And it will become convenient to site nuclear stations at sea, near the shore, and to pipe electricity to the coastal urban regions. This concept will become increasingly attractive as the price of coastal land soars and it becomes cheaper to site nuclear stations on floating platforms. Energy-intensive industries will probably also be established on these platforms around the nuclear stations.

We have seen that in the last decade or so of this century many very large nuclear plants will be installed throughout the world, and that energy-centres, the major contribution of nuclear energy to the development of the developing countries, will become conventional establishments. This nuclear future will have many international aspects and the role of international and regional organizations will be essential – political and national boundaries will often be too restrictive for the continued progress of nuclear technology and

Above: a model of an island to be built off the southern Californian coast for a dual purpose nuclear power plant supplying electric power and desalinated water to the coastal urban area

for the solution of many of the problems it will raise. The disposal of the huge amounts of radioactive wastes that will be produced, international regulations for the transport of radioactive material, and the standardization of nuclear law and insurance are some of the issues which will require international collaboration. And the high degree of co-operation stimulated between states by nuclear energy will assist peaceful relations between them.

Arsenals

Unfortunately, nuclear weapons technology is working in the opposite direction. In 1953, Eisenhower anticipated 'the day when fear of the atom will begin to disappear from the minds of people and the governments of east and west'. It is sad to relate that this day is not yet in sight. On the contrary, over the past quarter century, the destructive power of the world's arsenals of nuclear weapons has increased by leaps and bounds, and a sophisticated new theory of strategy – the strategy of deterrence – has even been built up around it.

The very first resolution of the United Nations General Assembly asked that nuclear weapons should be eliminated from national arsenals. Its complete failure rapidly led to an arms race between the United States and the Soviet Union. This race has spiralled upwards ever since, moving to new dimensions in 1952 with the development of thermonuclear weapons and in 1957 when it was demonstrated that thermonuclear warheads could be delivered accurately over almost limitless ranges by intercontinental ballistic missiles (ICBMs).

The realization that if one side attacked with its nuclear forces, the other would instantly retaliate in kind led to the perception of strategic nuclear deterrence. And in an effort to maintain this state of deterrence both superpowers have continually taken measures to ensure that their nuclear weapons and delivery systems are proof against possible counter-measures taken by the other side. But this continuous development of weapons systems has caused each superpower to arm itself to enormous levels of destructive capacity at a huge cost; it is estimated that between them they are now spending nearly $200 billion per year on military forces, a sum equal to about twice the Gross National Product of the United Kingdom.

The United States has so far deployed about 1,050 ICBMs in hardened underground silos and a fleet of 41 nuclear-powered ballistic missile submarines, each equipped with 16 polaris missiles. The land-based ICBMs include 550 Minuteman-I and 450 Minuteman-II solid-fuelled missiles, with a range of about 6,500 and 8,000 miles and warheads of 1 and 2 megatons respectively. Eventually, these will be replaced with Minuteman-IIIs, each equipped with a multiple independently-targetable re-entry vehicle (MIRV) carrying three 0.2 megaton warheads aimed at different targets which can be as far apart as 100 miles. The deployment of 500 Minuteman-IIIs, already begun, is due to be completed by 1974. About 50 Titan liquid-fuelled ICBMs are still in service; these have a range of 9,000 miles and carry a warhead of over 5 megatons but they have slower reaction times than do solid-fuelled missiles.

At ranges of about 7,000 miles ICBMs are sufficiently accurate to fall within about one mile of their targets,

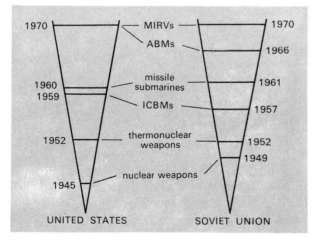

UNITED STATES   SOVIET UNION

The progression of the arms-race between the two superpowers. As time goes on the weapons become increasingly awesome, and since the beginning of the race there has not yet been any measured disarmament

an accuracy sufficient to ensure that a one megaton warhead would very severely damage a targeted city.

As if the ballistic missiles are not enough, the United States Strategic Air Command operates about 500 B-52 long-range (12,800 miles) bombers. A B-52 carries two 700-mile range air-to-surface missiles with thermonuclear warheads and is capable of carrying two more thermonuclear weapons in its bomb bays – a total explosive power exceeding that ever used in war.

Recently the 425-foot long US nuclear submarine, the *James Madison*, departed from the Groton, Connecticut shipyards to begin its sea-trials fitted with 16 Poseidon missiles equipped with MIRVs. At present, 13 of the nuclear submarines carry Polaris-II

Opposite: Minuteman take-off. By the end of this power-boost the missile must be accurately aligned on target as the remaining 95 per cent of its course is ballistic and uncontrolled. Left: a Minuteman en route to an underground silo. Bottom: a 20-kiloton fission bomb above a megaton thermonuclear weapon, and, to the right, an atomic artillery shell of 28cm calibre

195

missiles and 28 carry Polaris-IIIs. These missiles have
ranges of 1,700 and 2,800 miles respectively, but both
types carry 0.7 megaton warheads. And some Polaris-
IIIs have triple warheads.

It is planned to replace the Polaris missiles in 31 of
the nuclear submarines with Poseidons. And the *Madison* will be the first of these. A single Poseidon missile
is capable of delivering ten separate independently
targetable warheads, each with more than twice the
explosive power of the nuclear weapons dropped on
Japan. By 1975, when the last Poseidon-firing submarine takes to the oceans, a total of nearly 5,000 warheads will be deployable at sea – a weapon system
which will cost the Americans $18 billion.

As of now, the American strategic forces could
launch at least 4,000 thermonuclear warheads, each
with an average yield of over one megaton (about
1,000 from land-based ICBMs, 1,000 from sea-based
ICBMs and about 2,000 from aircraft). In addition,
there are all the nuclear weapons that tactical aircraft
can carry and the 7,000 tactical nuclear warheads in

Europe. By 1975, the number of strategic warheads deployed in ballistic missiles alone will have increased to 7,000.

The Soviet strategic forces include about 300 very large and 1,100 other ICBMs, having ranges between 5,000 and 10,000 miles, and about 750 intermediate range (about 1,000 miles) and medium range (2,000 miles) ballistic missiles. The Soviet Navy operates a large submarine fleet including about 60 nuclear-powered vessels, of which 50 can fire, on average, three ballistic missiles. A new type of nuclear-powered ballistic missile submarine, similar to the American Polaris, is being produced at the rate of at least four per year – each will carry 16 missiles for submerged firing, with warheads of about one megaton but comparatively short ranges of between 400 and 700 miles. But a new submarine-launched missile is being developed to carry a larger warhead over a longer range. The Soviet strategic air force has about 200 long-range bombers, with maximum ranges of about 8,000 miles,

With the advantage of great range and the mobility of its base the Polaris missile (below) has made land-based missiles obsolete. It can also be fired while the submarine is submerged

Part of the Soviet Union's display of formidable strategic missiles paraded each year through Red Square, Moscow

and about 50 of these carry one air-to-surface missile.

Both superpowers are deploying, or plan to deploy, anti-ballistic missile (ABM) systems designed to detect, identify, intercept and destroy the warheads of hostile missiles. The Soviet system, deployed around Moscow, is based on the solid-fuel Galosh ABM which is believed to carry a nuclear warhead of about 1.5 megatons and to have a range of a few hundred miles. And it is reported that a second-generation ABM system is under development, probably including a high acceleration, short-range missile. The American ABM system, called 'Safeguard', is being installed to protect ICBM sites and the first two installations are to be in North Dakota and Montana around sites that contain about one-third of the Minuteman force.

Safeguard, which will probably be expanded to 12 installations at a cost of not less than $8 billion, is based on the long-range Spartan missile with a warhead of over one megaton which can intercept an incoming missile at an altitude of about 200 miles at ranges of up to 400 miles from its location. The Sprint missile, designed to intercept enemy ICBMs which escape the Spartans, has a high acceleration – it can reach an altitude of about 12 miles in about 4 seconds – and carries a nuclear warhead equivalent to several tens of kilotons of TNT. A highly complex and expensive radar and computer system is associated with the American ABM system.

An ABM system is extremely complicated. For example, the computer 'brain', which collates information of the trajectories of the hostile missiles from all the radars and guides the ABMs to them, has a data handling and decision-making capacity greater than any previously constructed machine. Since none of the components have been fully tested under operational conditions it is not possible to assess the effective-

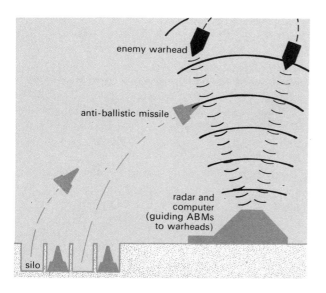

enemy warhead

anti-ballistic missile

radar and
computer
(guiding ABMs
to warheads)

silo

The principle by which anti-ballistic Missiles are deployed to intercept and hopefully to destroy enemy intercontinental ballistic missiles

ness of the system. But it can be predicted that an ABM system could be relatively easily overcome by the use of penetration aids, such as dummy missiles, or by suitable strategies like increasing the number of offensive missiles.

The British strategic nuclear force consists of four nuclear-powered submarines, each carrying 16 Polaris missiles with triple warheads. The French strategic force is based on 45 Mirage IV-A bombers, adapted for low-level penetration with 80-kiloton nuclear weapons. France is developing intermediate range (2,000 miles) ballistic missiles and plans to install 27 of these, in hardened silos in Provence, during the 1970s. The French navy has ordered four nuclear submarines, each armed with 16 missiles, scheduled for completion by 1975. China has demonstrated her technological capability by launching, into a high orbit, a 380-lb satellite. And she will probably soon deploy medium range missiles and ICBMs en route to a strategic nuclear force of superpower quality.

Unless the superpowers restrain the future development of nuclear weapons, new ones will undoubtedly emerge which will make even the awesome MIRV appear primitive. For example, the self-aligning boost and re-entry system (SABRE) is under development in the United States. This has a sensing mechanism to scan the ground below it and a computer to control the guidance system in making final course adjustments so that the warhead lands accurately on its target after manoeuvring against an ABM defence. And the Soviet fractional orbital bombardment system (FOBS) will be developed to an operational stage. These weapons are launched into a very low orbit around the earth about 100 miles high, and, at a point in the first orbit, a retro-rocket is fired to slow down the missile which then drops on to its target. The orbit would have an almost constant altitude, whereas an ICBM would follow a true ballistic trajectory which has much more curvature. Therefore, a FOBS would be undetected until it was within a range of about 900 miles – the warning time would be only about three minutes compared to ten minutes for an ICBM. And a missile on a fractional orbit trajectory could be made to approach a defender's radar system from a direction which would make its detection difficult.

New defensive weapons will also be developed. A high acceleration (500g) anti-missile and sea-based ABMs (SABMIS) are under development in the United States. A future ABM system might be based upon a defensive screen established by exploding very large nuclear warheads in space. If this occurred inside the earth's magnetic field, the charged particles released by the explosions would move along the lines of the field and might achieve a density sufficient to inactivate an incoming enemy warhead. Defensive screens of small pellets and gases have been suggested as alternatives to charged particles.

Proliferation of weapons
Because nuclear reactors can be used to produce fissile material for the production of nuclear weapons, the spread of nuclear power to many countries could lead to a considerable increase in the number of nuclear-weapon states, a prospect which causes much concern. But, because of the great benefits to be derived from the use of nuclear energy, the spread of power reactors to any country which needs them should not, and probably could not, be hindered for this reason.

If power reactors are operated for the production of the cheapest electricity, the plutonium they produce is not suitable for immediate use as the fissile material for efficient nuclear weapons because of the presence of non-fissile isotopes of plutonium, particularly Pu-240. Weapons-grade plutonium should contain no more than 10 per cent of these isotopes, and preferably less. But nuclear power reactors can be operated in such a way that the plutonium produced in them is sufficiently pure in Pu-239 for use in nuclear weapons. This can be achieved by limiting the burn-up of the uranium fuel to values of less than 1,000 MWD per tonne of uranium. Under normal operating conditions, the fuel burn-up in power reactors is typically an order of magnitude higher than this value and the plutonium recovered from the spent fuel elements has a Pu-239 content of between 60 and 70 per cent. The Magnox reactors are an exception because they normally operate with burn-up values of about 3,500 MWD per tonne; the Pu-239 content is about 85 per cent.

But some of the plutonium normally produced in power reactors might be usable in a very primitive nuclear weapon. A relatively large amount of this plutonium would be required and consequently the physical size of the weapon would be great. And there would be the danger that it would over-heat due to the spontaneous fission of Pu-240. But it is possible that a method will eventually be developed for the enrichment of Pu-239 in the plutonium produced by power reactors to levels which would make it suitable for nuclear weapons.

The potential of countries with nuclear power programmes to produce plutonium is increasing rapidly. By 1975, the non-nuclear weapon states will be producing about 10,000 kilograms per year. If this were weapons-grade, it would be sufficient to produce about 2,000 nuclear weapons per year. And by 1980, total world production of plutonium will have reached 100,000 kilograms per year.

If a country accumulated weapons-grade plutonium from its reactors and wished to acquire an effective nuclear force it would need sufficient scientific, industrial, and technical capability to assemble and test nuclear warheads and to develop delivery vehicles, whether missiles or aircraft. This would involve considerable expenditure, both financial and manpower. Trained personnel represented by scientists – physicists, chemists, metallurgists, mathematicians, engineers – and skilled workers – machine-tool operators, electricians, pipefitters, welders, sheet-metal workers, furnace operators, chemical plant operators and instrument makers – would be required to manufacture and assemble components to precise specifications. And, for the indigenous development of a nuclear force, well-established aircraft, missile and electronic industries would be necessary. Although some types of military bombers able to carry nuclear weapons – such as the British Canberra, the American B-51 and the French Mirage IV – can be purchased abroad, sophisticated missiles cannot.

It has been estimated that the cost of a modest nuclear capacity consisting of a force of 30 to 50 jet bombers, 50 medium-range (3,000 kilometres) missiles in soft emplacements and 100 plutonium warheads would be about $170 million per year spread over ten years. A small but high-quality nuclear force acquired in two stages, each of five years duration, comprising 10–15 bombers and 15–20 nuclear weapons at the end of the first stage and extended to include twenty to thirty thermonuclear weapons, 100 intermediate range missiles and two missile-launching submarines during the second stage, would cost a total of about $560 million per year over ten years. For comparison, the costs of the French and British military nuclear programmes have, so far, each been about $9 billion. But the cost of developing simple nuclear warheads is steadily decreasing as the technology is increasingly becoming public knowledge.

Mirage IV strike bomber – at present the backbone of the French strategic nuclear force and capable of delivering nuclear weapons of up to 60-kilotons

Forty-four countries have defence budgets greater than the $170 million per year estimated for a modest nuclear force and 18 of them have defence budgets greater than the $560 million per year estimated for a small high-quality nuclear force. But no country would devote all of its defence expenditure to this purpose and only 16 countries have such large defence expenditures that the cost of nuclear forces would represent a relatively small component of the total defence budget. Excluding the existing nuclear-weapon powers these are: the German Federal Republic, Poland, Italy, the German Democratic Republic, Canada, Czechoslovakia, India, Japan, Australia, Sweden and the Netherlands. In addition, Israel, the United Arab Republic, North Korea, Yugoslavia, Spain, Rumania, Pakistan and Belgium, have defence budgets greater than three times the cost of the modest nuclear force, and these countries could therefore just about contemplate this expenditure without reallocating major resources from non-military activities. But the economies of many countries are growing rapidly and, if the percentage of the Gross National Product devoted to defence remains constant, the absolute amount of money available for weapons systems will increase at a corresponding rate. If the defence-spending ratio of Japan, for example, is maintained then, assuming that the economy grows as rapidly as is predicted, the defence budget will increase from its present $1.4 billion to about $2.4 billion in 1974 and to about $5 billion by 1980. The number of countries able to afford nuclear forces will therefore increase.

Motives for acquiring nuclear arms

Each of the present nuclear-weapon powers has described its motives for acquiring a nuclear arsenal as defensive. But it is clear that the possession of nuclear weapons has a different significance to each, both in terms of military power and political security. If any of the present non-nuclear-weapon powers decide to acquire nuclear weapons it will be for a complex set of motives which will differ in each case. An argument sometimes used for the acquisition of nuclear weapons is that they enhance a country's influence in international affairs to an ensuing increase in prestige. A second argument is that nuclear weapons, by their deterrent value, increase a country's security. A third possible motive is the perception that nuclear weapons increase political independence. A fourth possibility is that the failure of the existing nuclear-weapon states to limit the development and deployment of their nuclear weapon systems may lead to the acquisition of these weapons by other states.

The view that nuclear weapons somehow enhance the status of a power over the long term cannot be supported by their available evidence. There may be an imponderable element of prestige in the demonstration of technological prowess indicated by the development of a nuclear force, but this does not last for long. Britain, for example, is generally reckoned to have lost status in the period since she acquired nuclear weapons. However, status could be a significant factor in the calculations of the German Federal Republic, Italy and Japan. Japan may feel that, as the third greatest industrial state, she should possess nuclear weapons to retain her influence in Asia, which may be jeopardized by the rapidly rising influence of China. The continuing importance placed by the United Kingdom and France on their nuclear status could cause the German Federal Republic and, to a lesser extent, Italy to feel the need for this status to gain recognition of their position in Europe. India, on the other hand, may be more influenced by her perceptions of a threat from China. And Sweden may, in the future, argue that she needs nuclear weapons to protect her neutrality.

It is generally believed that any further increase in the number of nuclear-weapon states will lead to greater tension and a more unstable world. The treaty on the non-proliferation of nuclear weapons is designed to prevent this from happening. It prohibits non-nuclear-weapon states from receiving, manufacturing, or otherwise acquiring nuclear weapons and nuclear-weapon states from transferring, or in any

With an energy source of greater potential than any hitherto encountered by man we can destroy all life on earth in the way that this tumour ridden twig has ben stunted and killed and we can use our machines, like the generator opposite – a part of the nuclear research apparatus – to discover how to handle the energy of the atom to add value to all life

way assisting, other states to acquire these weapons. But the treaty also obligates the nuclear-weapon states to negotiate 'effective measures relating to the cessation of the nuclear arms race at an early date'. Its long-term viability will largely depend on whether this obligation is taken seriously. The treaty should then discourage near-nuclear-weapon signatories from acquiring nuclear weapons for reasons of status since any gain in status would be outweighed by the loss of prestige which would follow the violation of an international agreement.

But several important non-nuclear-weapon countries – like India, Brazil, Argentina and South Africa – have not joined the treaty. Furthermore, on present indications, it is unlikely that the nuclear-weapon states will achieve sufficiently effective disarmament measures to satisfy some of the signatories. The treaty is, therefore, a fragile instrument and cannot be relied upon to prevent the proliferation for all time.

The United Nations Report on the 'Effects of the Possible Use of Nuclear Weapons and on the Security and Economic Implications for States of the Acquisition and Further Development of these Weapons' concluded that if nuclear and thermonuclear weapons were ever used 'in numbers, hundreds of millions of people might be killed, and civilization as we know it, as well as organized community life, would inevitably come to an end in the countries involved in the conflict. Many of those who survived the immediate destruction, as well as others in countries outside the area of conflict, would be exposed to widely-spreading radioactive contamination, and would suffer from long-term effects of irradiation and transmit, to their offspring, a genetic burden which would become manifest in the disabilities of later generations.'

At the very beginning of the nuclear age the tragedies of Hiroshima and Nagasaki brought destruction and human suffering on a scale hitherto unknown. Will mankind end the nuclear age by repeating this folly on a truly cataclysmic scale, or will the good that nuclear energy can do be allowed to predominate?

GLOSSARY

alpha particle: particle identical to the nucleus of the helium atom consisting of two protons and two neutrons bound together, emitted in radioactive decay.

atomic number: number of protons in the nucleus of the atom.

beta particle: electron emitted from the nucleus of an atom in certain types of radioactive decay.

blanket: fertile material put round a reactor core to breed new fuel.

body burden: total amount of radioactive material present in the body at any time.

breeder reactor: nuclear reactor that also generates nuclear fuel.

burn up: fraction of atoms in a reactor fuel which has undergone fission. Also the total amount of heat released per unit mass of fuel, usually expressed in megawatt-days per tonne.

chain reaction: process in which one nuclear event sets up conditions which permit the same event to take place in another atom.

control rods: rods, plates or tubes of steel or aluminium, containing a strong absorber of neutrons, used to maintain a reactor at a given power level.

critical: condition in which a chain reaction is maintained at a steady rate; just self-sustaining.

curie: unit of radioactivity; the radioactivity emitted from the quantity of an isotope that decays with 37 thousand-million disintegrations per second. This quantity depends on the half-life of the substance. Less than a thousandth of an ounce of strontium-90 gives one curie and one tonne of natural uranium would give the same.

decay: gradual decrease in radioactivity of a radioactive substance due to nuclear disintegration; what remains is a different element.

disintegration: process in which a nucleus spontaneously emits one or more particles, normally alpha or beta particles and gamma rays.

fall out: radioactive dust and other matter falling back to the earth's surface from the atmosphere after a nuclear explosion.

fast reactor: a nuclear reactor in which most of the fissions are caused by neutrons moving with the high speeds they possess at the time of their birth in fission. Such reactors contain little or no moderator.

fertile material: isotopes capable of being transformed into fissile material by the absorption of neutrons, typically in a reactor.

fissile (or fissionable): capable of undergoing fission when hit by a slow neutron.

fission: splitting of a heavy nucleus into two (or very rarely more) approximately equal fragments – the fission products; fission can be spontaneous or caused by the impact of a neutron.

fusion: process of building up more complex nuclei by the combination of simpler ones; usually accompanied by release of energy.

half-life: time taken for a radionuclide to decay to half its initial value; time taken for half of the atoms to decay; may vary from less than a millionth of a second to millions of years.

heavy water: water consisting of molecules in which the hydrogen is replaced by deuterium; present in about 1 part in 5000 in water.

ion: atom or molecule that has lost or gained one or more electrons – it is therefore electrically charged.

isotopes: nuclides having the same atomic numbers but different numbers of neutrons.

leukaemia: disease of the blood, corresponding to cancer in tissue, in which there is an excess of white corpuscles; can be induced by excessive exposure to radiation.

magnox: magnesium alloy used to can the uranium in some reactor fuel elements.

multiplication factor: in a nuclear reactor, the ratio of the number of neutrons in consecutive generations of the fission process.

neutron: one of the two particles of the nucleus; has no electric charge and is slightly heavier than a proton.

nuclide: species of atom characterized by the number of protons and the number of neutrons in its nucleus.

plasma: very hot gas consisting mostly of positive ions and electrons in nearly equal concentrations.

plutonium: element produced by the neutron bombardment of uranium-238. Plutonium-239 is a fissile material; has awkward physical properties and is very poisonous.

proton: one of the two particles of which nuclei consist; has an equal but opposite electrical charge to that of an electron but has a mass 1,837 times as great.

radioisotope: radioactive nuclide of a given element.

safety rod: neutron-absorbing rod which, in an emergency, can quickly be dropped into a reactor to shut it down.

thermonuclear reaction: nuclear fusion reaction brought about by temperatures, which must exceed 20 million degrees to sustain a reaction.

tracer: small quantity of a radionuclide which can be added to normal matter so that by subsequent analysis the presence and position of the normal matter can be detected.

transuranic elements: artificial elements which have heavier and more complex nuclei than uranium; can be made by neutron bombardment of uranium.

uranium: heavy metal. Uranium-235 is the only naturally occurring readily fissile isotope, uranium-238 is a fertile material, uranium-233 is a fissile material that can be produced from thorium-232 by neutron absorption. Natural uranium contains 1 part in 140 of uranium-235.

BIBLIOGRAPHY

Introduction

Glasstone, S., *The Effects of Nuclear Weapons*, Washington, 1962.

Lifton, R. J., *Death in Life: Survivors of Hiroshima*, New York, 1968.

Morris, C., *The Day they Lost the H-bomb*, New York, 1966.

Stonier, T., *Nuclear Disaster*, New York, 1964.

1 The Birth of the Bomb

Davis, N. P., *Lawrence and Oppenheimer*, London, 1968.

Glasstone, S., *Sourcebook on Atomic Energy*, Princeton, 1958.

Groeff, S., *Manhattan Project*, London, 1968.

Jungk, R., *Brighter than a Thousand Suns*, London, 1965.

Lamont, L., *Day of Trinity*, New York, 1965.

Smyth, H. D., *Atomic Energy for Military Purposes*, Washington, 1945.

Moss, N., *Men who Play God*, New York, 1969.

2 Atoms for Peace

Hogerton, J. F., The Arrival of Nuclear Power, *Scientific American*, February, 1968.

Jay, K. E. B., *Calder Hall*, London, 1956.

Jensen, W. J., *Nuclear Power*, Cambridge, 1969.

Loftness, R. L., *Nuclear Power Plants*, Princeton, 1964.

Medical Research Council, *The Hazards to Man of Nuclear and Allied Radiations*, London, 1960.

3 Nuclear Realities

Burn, D., *The Political Economy of Nuclear Power*, London, 1967.

Calder, D., *Living with the Atom*, Chicago, 1962.

Curtis, R. and Hogan, H., *Perils of the Peaceful Atom*, London, 1970.

International Atomic Energy Agency, *Nuclear Energy Costs and Development*, Vienna, 1970.

Mullenbach, P., *Civilian Nuclear Power; Economic Issues and Policy Formation*, New York, 1963.

Schubert, J. and Lapp, R. E., *Radiation; What it is and what it does*, New York, 1957.

Thompson, T. J. and Beckerley, J. G., *The Technology of Nuclear Reactor Safety*, Cambridge, U.S.A., 1964.

Weinberg, A. M., Breeder Reactors, *Scientific American*, January, 1960.

4 The Versatile Radionuclide

Barnaby, C. F., *Radionuclides in Medicine*, London, 1969.

International Atomic Energy Agency, *Induced Mutations in Plants*, Vienna, 1969.

International Atomic Energy Agency, *Sterile-male Technique for the Eradication or Control of Harmful Insects*, Vienna, 1969.

Mander, J., *Atoms at Work*, Holywood, 1957.

5 Mini-reactors

Daugherty, C. M., *City Under the Ice: the Story of Camp Century*, New York, 1963.

Fox, C. H., Packaged Nuclear Reactors, *New Scientist*, May 17, 1962.

Kramer, A. W., *Nuclear Propulsion for Merchant Ships*, Washington, 1962.

Nuclear Energy in Space, fourteen articles in *Nucleonics*, April, 1961.

6 Peaceful Nuclear Explosions

Teller, E. et al., *The Constructive Uses of Nuclear Explosives*, New York, 1968.

Proceedings of the Third Plowshare Symposium, *Engineering with Nuclear Explosives*, Springfield, 1964.

7 The Nuclear Future

Barnaby, C. F., *The Nuclear Future*, London, 1969.

Beaton, L., *Must the Bomb Spread?* London, 1966.

Chayes, A. and Weisner, J. B. (Eds.) *An Evaluation of the Decision to Deploy an Anti-ballistic Missile System*, New York, 1969.

Hammond, R. P., Low Cost Energy: A New Dimension, *Science Journal*, January 1969.

Jones, R. E., *Nuclear Deterrence*, London, 1968.

Sherman, M. E., *Nuclear Proliferation: the Treaty and After*, Toronto, 1968.

United Nations, *Effects of the Possible Use of Nuclear Weapons and the Security and Economic Implications for States of the Acquisition and Further Development of these Weapons*, New York, 1968.

SOURCES OF ILLUSTRATIONS

213

The diagrams were drawn by Sandra Barnaby.

INDEX